Tagungsband des 2. Kongresses Montage Handhabung Industrieroboter

Thorsten Schüppstuhl · Jörg Franke
Kirsten Tracht
(Hrsg.)

Tagungsband des 2. Kongresses Montage Handhabung Industrieroboter

Springer Vieweg

Herausgeber
Thorsten Schüppstuhl
Hamburg, Deutschland

Kirsten Tracht
Bremen, Deutschland

Jörg Franke
Erlangen, Deutschland

ISBN 978-3-662-54440-2 ISBN 978-3-662-54441-9 (eBook)
DOI 10.1007/978-3-662-54441-9

Die Deutsche Nationalbibliothek verzeichnet diese Publikation in der Deutschen Nationalbibliografie; detaillierte bibliografische Daten sind im Internet über http://dnb.d-nb.de abrufbar.

Springer Vieweg

Gedruckt auf säurefreiem und chlorfrei gebleichtem Papier

Springer Vieweg ist Teil von Springer Nature
Die eingetragene Gesellschaft ist Springer-Verlag GmbH Deutschland
Die Anschrift der Gesellschaft ist: Heidelberger Platz 3, 14197 Berlin, Germany

Inhaltsverzeichnis

Development of a flexible special kinematics for the fully mechanized production of tubular nodes for offshore foundation structures

Hagen Herholz[1,a], Steffen Dryba[1,b], Mirko Gründler[1,c], Martin-Christoph Wanner[1,d]

[1]Fraunhofer Research Institution Large Structures in Production Engineering, Albert-Einstein-Straße 30, 18059 Rostock, Germany
[a]hagen.herholz@hro.ipa.fraunhofer.de,
[b]steffen.dryba@hro.ipa.fraunhofer.de,
[c]mirko.gruendler@hro.ipa.fraunhofer.de,
[d]martin-christoph.wanner@hro.ipa.fraunhofer.de

Abstract This article is about an innovative welding system for the fully mechanized production of large dimensioned tubular nodes which are designed as welding constructions. The system is characterized by its flexibility regarding to the weight of the components as well as the type of installation. First, the main assemblies of the robot welding system and their essential components are described. Problems concerning the welding process as a consequence of the design concept are presented. Additionally, solution approaches to these problems are discussed on the basis of experimental results. Finally, a description of potential process steps within the production of tubular nodes, which also could be implemented by the use of the described manipulator, is given.

Keywords: tubular nodes, offshore foundation structure, multi-layer welding, GMA-tandem welding, orbital welding system, flexible kinematics

1 Introduction

At the moment, the offshore wind energy market is strongly growing. One essential reason for this trend is the higher wind speed as well as almost constant wind conditions. In this way, higher electricity yields can be achieved by wind energy compared to wind turbines on mainland [1]. Furthermore, these development trends are supported by the energy policy of the German government. This includes the installation of wind turbines with a total power of 15000 MW in the Baltic and North Sea till 2030 [2].

The necessary foundation structures for wind turbines or offshore substations for wind farms are mainly realized as monopiles. However, other foundation structures like tripods or jacket structures have more preferable dynamic stress resistances, less construction weights and lower material costs. This fact is based on the higher degree of automation in the serial production of monopiles. In order to increase the competitiveness of the structurally more complex, but considerably more efficient, foundation structures, a transfer of the predominantly manual production towards an industrial serial production is necessary.

The main focus is the automated production of the tubular nodes. These tube to tube connections, which are designed as welded constructions, occur in large numbers, especially in jacket structures (**Fig. 1**). Due to the high workpiece mass (partly > 10 t per node) and the technologically challenging welding task, these nodes are mostly welded manually in a time-consuming production step.

Fig. 1 Offshore substation wind farm Borkum Riffgrund [3]

Economical increase concerning the level of mechanization in production of the tubular nodes is currently being countered by several factors. Especially the diversity of occurring node types and the production tolerances, required by the classification societies (e.g. DNV-GL) have to be noted. The large component diversity results from load- and installation-optimized design concepts of the foundation structures. This leads to a lack of standardization of the tubular node geometry and in this way to a higher demand for flexibility of the production systems.

2 State of the Art / Related Work

There were many efforts to develop automation solutions for the production process of tubular nodes in the past. One possible strategy to achieve a technological optimization of the welding, and thus the production process, is to manipulate the processed workpiece (tubular node).

In this case, the preassembled tubular node (connection of the structural components by a root weld) is manipulated for realizing the welding process in optimal working position 1G (flat position). This is achieved by using a rotary and tilting positioner to clamp, position and orientate the workpiece [4] (**Fig. 2**).

Fig. 2 Rotary and tilting positioner with tubular node [4]

The manipulation of the welding torch is accomplished by a six-axis vertically articulated robot arm connected with a three-axis gantry. One advantage of the system concept is the exclusive achievement of the welding work in flat position (1G). In this way, the influence of gravity on the molten pool is minimalized and an increase of the deposition rate is possible. In contrast, the limited accessibility of the joint during a set-up of several brace stubs on the chord can (complex node geometry) is a problem. Furthermore, there are restrictions of the allowed total mass (approx. limited to 10 t) regarding the manipulation of the working pieces [4].

One possibility to avoid the manipulation of work pieces is to use a rail guided welding system, as explained in [5]. This system consists of three-dimensional curved rail elements having a supporting arm with magnetic holders and a variable number of connected motion modules. The main function of the motion module is to fix the necessary hardware for the welding process (welding torch, wire feed system, drive unit etc.) [5] (**Fig. 3**, l.)

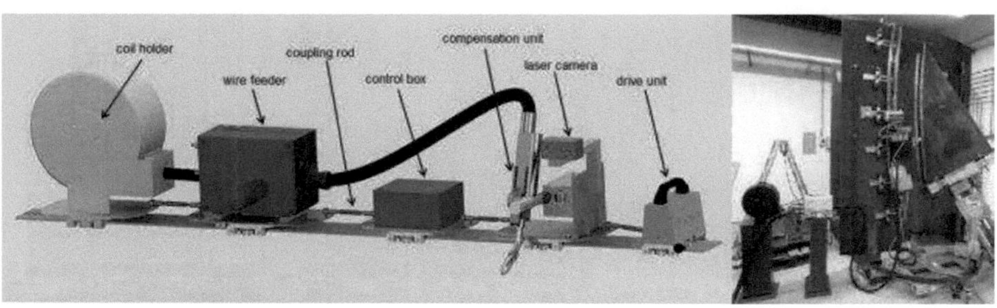

Fig. 3 Rail guided welding system (l.), Test stand (r.) [5]

The necessary data for realization of a multi-layer welding (positioning and orientation of welding torch) is determined by measuring the joint geometry with a two-dimensional laser scanner. This data will be assigned to the motion programs. Advantages of such a fully mechanized welding system are the lower total mass of the system technology and the flexibility of the number of motion modules. Disadvantages consist of the mandatory adaption of the rail elements to the work piece geometry and the limited accessibility of small connection angles of the brace stub. The dimensions of the on-board equipment technology contribute to the limited accessibility. Additionally, the welding power source and the shielding gas supply have to be located separately from the manipulator. This further complicates the handling of the energy and media supply for work piece geometries used in the offshore area (**Fig. 3**, r.).

3 Description of system concept and main assemblies

Within a research project from 2014 to 2016, a welding system for the main layer of tubular nodes was developed. The system is especially well suited to deal with large workpiece dimensions in the offshore industry. The concept is to connect a cylindrical coordinate manipulator to the tubular node, which is to be joined by using form or force closure. Moreover, small and flexible industrial robots are moved along the welding seam.

In contrast to the solution proposed in [5], the energy, media and data supply is to be provided by a special supply platform. This platform is synchronized with the robot movement, and thus opens the possibility for an endless rotary movement of the welding manipulator as well as the system periphery.

Fig. 4 Orbital welding system (commissioning scenario)(l.), concept for portal installation (r.)

The resulting system (**Fig. 4**, l.) has a modular design and three main function modules. The basis for the manipulator is a clamping mechanism (**Fig. 5**, l.) [6], which can be lowered from the supply platform via three cable winches and inserted into the tube supports. The module is positioned on the edge of the tube and is locking the translational degree of freedom in the axial direction of the tube by form-closure. A clamping device with four gripping yaws is used for centering and also enables a connection via force closure. This gripper is mechanically actuated by a ball screw assembly and can be used for tubes of different diameters. Moreover, the gripper compensates changes during the welding process with a prestressed disk spring set, e.g. the widening of the tube due to thermal expansion.

Fig. 5 Clamping mechanism (li.), Cylindrical coordinates manipulator (m.), Supply platform (r.)

The system is followed by the cylindrical coordinate manipulator, which consists of a rotary and translational joint (**Fig. 5**, m.). The rotary axis is implemented by a ball bearing slewing ring with external teeth on which one or more rigid linear axes can be mounted. A configuration with two translational axes and an offset of 180 degree is optimal to minimize welding distortion and loads such as holding torques for the drives. It is possible to work with up to four robots at the same time due to the modular design. Control, media and power supply for the robots and the welding equipment are guided through the linear axes upwards to the supply platform (**Fig. 5**, r.). It contains the controller of the robots and further drives, as well as welding power source and the filler wire supply. The energy support, the earth cable of the welding system, the shielding gas supply and a bus system for control are slip-ring-system-guided. This system has a hollow shaft where the static supply conductors are guided such as the support of the preheat technology and the drive of the clamp mechanism. Furthermore, there are three cable winches on the platform for moving down the system on the pipe socket. An installation of the supply platform is possible on a fixed or a mobile base (**Fig. 4**, r.), which is advantageous because of the possibility to transfer the entire system technology within one or more production sites (also overseas).

4 Preparation of the robot application

The programming of the robot system is a challenging task. Hence, it is necessary to do it offline and sensor-based. To solve this problem, the CAD/CAM-Software "Siemens Process Simulate" is used to create the robot paths. This ensures that there are no collisions between robot and workpiece, that the robot is capable to reach the desired position in terms of axis limitations and to convert it into executable robot programs by using a feasible interpreter. In addition to the position specification, all welding parameters are transferred, too. A great advantage of such an open, offline programming system is the option of incorporating custom libraries and algorithms into the system and creating application-specific graphical user interfaces. The proper move of a welding torch, in particular in multi-wire welding processes, requires the definition of all six degrees of freedom. The whole robot system consists of a cylinder coordinate manipulator coupled with an articulated robot with six joints and because of that the system is kinematic redundant. The control of the industrial robot used here (Kuka KR10 R900 sixx C) provides a function for this purpose called "mathematical coupling", which allows the synchronized operation of all eight axes with simultaneous continuous path motion of the end effector. Furthermore, the geometrical tolerances and the seam preparation of the tube connection are problematic areas. For this reason, it is not possible to assume an ideal shape geometry of the working piece. Therefore, the concept for the referencing and programming sequence consists of the following steps, which have to be done for each welding layer.

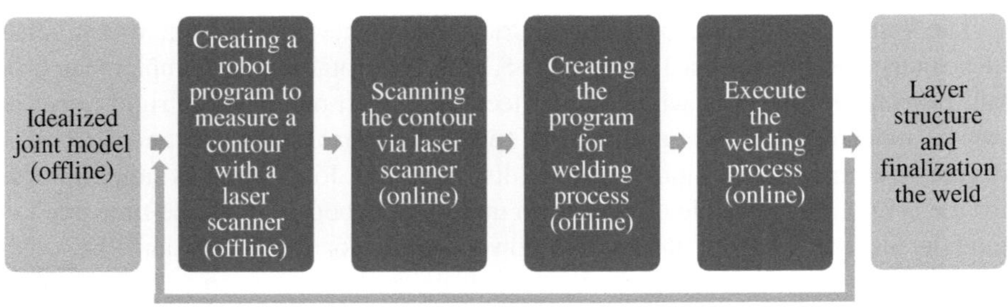

Fig. 6 Procedure of offline programming

In the first step an ideal model of the tubular node is created by the use of predefined geometrical parameters. Since the welding manipulator is mechanically connected to the tube, it is possible to predetermine reference points on the path at a predefined distance along the intersection contour of the two tubes, based solely on the available information (**Fig. 7**). The aim of the first measurement run is to obtain the actual geometrical parameters of the seam preparation and the filling level of the root welding. By using these results, the layer planning for the entire welding seam and the offline path planning of the robot motion is executed.

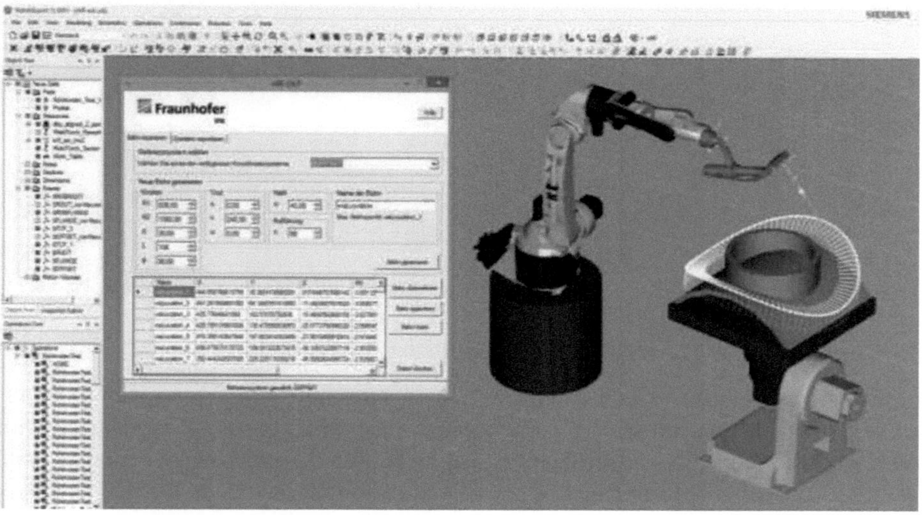

Fig. 7 Interface and offline path planning for contour measurement (Siemens Process Simulate)

5 Welding technology

The developed orbital welding system requires the procedure of welding work in the welding positions 1G (flat position), 2G (horizontal Position) and 3G (vertical up and vertical down position). The working positions go along the intersection line from brace stub to chord can and directly merge into another. Furthermore, varying seam inclination angles arise in both 3G positions. The range of those angles depends directly on the joint partners.

For the determination of suitable welding parameters regarding the main working positions, welding tests were carried out on T-joints ($t_{Webplate}$ = 30 mm, $t_{Flangeplate}$ = 60 mm) in offshore steel quality S355 G10+M with a constant groove angle of 45° (HV seam preparation) (**Fig. 8**, l.). A 90° tubular node connection with a diameter ratio of 2 : 1 (chord can : brace stub) was used for determination of seam inclination angles in the corresponding working positions. As a manipulator, a six–axis vertically articulated arm robot (guiding of welding torch) combined with a rotary and tilting positioner (orientation of test specimen), was used (**Fig. 8**, r.). The welding tests were carried out with GMA-tandem system (pulse arc) using a metal-cored wire (Ø = 1,2 mm, type T46 6 MM 1 H5) under shield gas M21 (82 % Ar, 18 % CO_2).

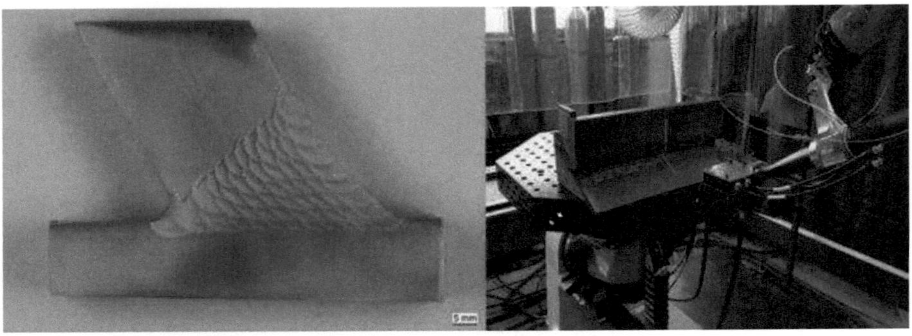

Fig. 8 Macro section of test specimen (l.), Welding test stand (r.)

The primary objective of T-joint tests was to determine a position-spanning applicable parameter set with maximum deposition rate, while ensuring compliance with minimum requirements for mechanical and technological properties of tubular node welding connections, formulated by the classification society DNV-GL in [7]. In order to demonstrate the process capability of the GMA-tandem process with metal-cored wire, the welded specimens were tested using destructive and non-destructive methods on basis of [7] under supervision of DNV-GL and in cooperation with SLV Mecklenburg-Vorpommern GmbH. In consequence of a general fulfillment of requirements, an AIP (Approval in Principle) for the orbital welding system as well as for the welding strategy was issued by DNV-GL. The AIP is the prerequisite for the transfer of the results from research and development activities into industrial practice.

Within the scope of experiments for the determination of a basic parameter set for metal-cored wires, it was found that the maximum deposition rate (MR) is substantially limited by welding position 3G (vertical up position). Considering the described test conditions, a value of MR = 5,9 kg/h was achieved. The limit of process stability was manifested in exceeding a critical wire feed rate in combination with an underrun of a critical feed rate. This led to a flowing weld metal against the feed direction (**Fig. 9**), which might be the consequence of the relatively large melting pool superimposed by gravity influence (inclination angle).

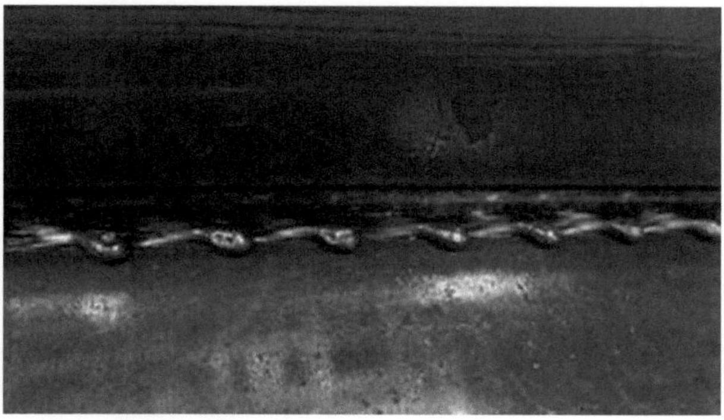

Fig. 9 Flowing weld metal

In order to investigate suitability as well as economic potentials of alternative wire electrodes, further experiments were carried out using solid wires (Ø = 1,2 mm, type G 46 6 M21 3Ni1). The test conditions for the determination of a position-spanning parameter set with a maximum deposition rate were maintained analogous to tests with metal-cored wires. Starting with vertical up position (critical position), the deposition rate was increased up to the limit of process stability. As a result, a deposition rate of MR = 9,1 kg/h was achieved. Subsequently, the suitability of the determined parameter set was validated in the remaining welding positions (1G, 2G, 3G down). The increase of deposition rate by approx. 54 % compared to metal-cored wires, possibly result from the axial drop transfer of solid wires (diffuse drop transfer of metal-cored wires). Furthermore, the cone-shaped arc (bell-shaped arc of metal-cored wires) may have a positive influence on the melting pool support effect [8]. With regard to the technical and technological boundary conditions of the investigations, the usage of solid wires is economically expedient.

Further research focuses on technically optimized welding parameters for the production of multilayer welding. In order to reduce the welding start points and to maximize the layer quality an alternative concept for circular bevel groove weld seam preparation along a joint contour was developed. Depending on a determined groove angle at the Crown Point of the cutting contour, the objective was to receive a constant cross section as a function of a varying groove angle. The controlled influence of weld bead geometry is necessary to create a constant degree of filling and for further application of the weld seam preparation concept. Significant deviations considering the geometry of the welds appear due to the varying inclination angles by the use of the same welding parameters in any welding position. One part of the results for the main welding positions is shown in Table 1 (Reference parameter).

An impact on the weld bead is possible by adapting peripheral parameters, such as welding torch orientation and arc-length-correction (**Table 1**, Parameter adjustment), whereby other basic welding parameters do not have to be changed. As a consequence, a constant degree of filling (multi-layer welding) of the seam cross-section is possible. Extensive series of tests have also been used to investigate parameter adjustments in intermediate positions (e.g. for occurring inclination angles in the region between 3G up and 1G. These tests provided similar positive results.

Table 1 Weld bead geometry in the main welding positions

Reference parameter (initial situation)			Parameter adjustment (optimization)		
Position/ experiment no.	Weld bead height [mm]	Weld bead width [mm]	Position experiment no.	Weld bead height [mm]	Weld bead width [mm]
1G / 20.0	5,67	5,17	1G / 33.-7	5,35	4,48
2G / 10.0	5,92	4,05	2G / 10.-7	5,13	4,27
3G up / 15.0	4,42	4,05	3G up / 20.-5	5,08	4,51
3G down / 20.0	6,28	5,09	3G down / 20.-5	5,36	4,52
$\bar{x}$ (average)	5,57	4,59	$\bar{x}$ (average)	5,22	4,45
Upper deviation	0,71	0,58	Upper deviation	0,14	0,08
Lower deviation	1,15	0,54	Lower deviation	0,18	0,17

Future work in the area of welding technology consists of the continuous build-up of a welding parameter database. The aim is to reference the parameter sets on basis of geometrical information of the tubular node connection and to transfer them to the offline programming.

6 Summary and Outlook

The orbital welding system presented in this paper has been successfully built up, put into operation and patented [9]. Fundamental welding research has been carried out by experimental studies. A transfer of the fully mechanized solution towards an automated system is possible depending on the type of installation (integration of workpiece feeding is necessary). Current work is focussed on transferring the results of welding parameter studies to the production of real components using the orbital welding system. Furthermore, the suitability of the described system is currently investigated regarding alternative processing tasks such as contour cuts on pipes or post-processing of tubular nodes (e.g. grinding of the seam transitions).

References

1. Gasch, R.; Twele, J.; Windkraftanlagen: Grundlagen, Entwurf, Planung und Betrieb; Springer Vieweg; Wiesbaden 2016

2. Bundesministerium für Wirtschaft und Energie: Erneuerbare Energien auf einen Blick, Information online on http://www.bmwi.de/DE/Themen/Energie/Erneuerbare-Energien/erneuerbare-energien-auf-einen-blick.html (19.12.2016)
3. Bladt Industries A/S, Information online on http://www.bladt.dk/allphotos/offshore-substations.aspx (19.12.2016)
4. ST3 Offshore Sp. Z o. o., Information online on https://st3-offshore.com/production/equipment/ (19.12.2016)
5. Mückenheim, U.; Sondershausen, C.; Ströfer, M; Keitel, S; Müglitz, J.: Vollmechanisierte Herstellung von dreidimensional gekrümmten Schweißverbindungen; DVS-Berichte Band 317; Düsseldorf 2015
6. Hesse, S.; Monkmann, G.; Steinmann, R.; Schunk, H.; Robotergreifer; Hanser Verlag; München Wien 2004
7. Germanischer Lloyd SE: Rules for the Certification and Construction, IV Industrial Services, Part 7 Offshore Substations; Hamburg 2013
8. Metting, G.: Fülldrahtschweißen von Dickblechen, eine wirtschaftliche Alternative; voestalpine Böhler Welding; DVS BV Gelsenkirchen; 2015
9. Wanner, M.C.; Dryba, S., Weidemann, B., Harmel, A.; Schweißanordnung zum dauerhaften Fügen eines ersten rohrförmigen Bauteils mit einem zweiten Bauteil; Fraunhofer Gesellschaft; 2016; DE 10 2015 206 044 A1

Enabling Domain Experts to Model and Execute Tasks in Flexible Human-Robot Teams

Dominik Riedelbauch, Tobias Werner, and Dominik Henrich

Chair for Robotics and Embedded Systems,
Universität Bayreuth, D-95440 Bayreuth, Germany,
dominik.riedelbauch@uni-bayreuth.de,
http://robotics.uni-bayreuth.de

Abstract. Recent advances in safe human-robot coexistence make collaboration of humans and robots in achieving common goals feasible. We propose a concept that treats human and robot agents as equal partners in executing a task specified by a shared task model. Equality between agents offers high flexibility, as e.g. the team composition may change arbitrarily without interrupting the working progress. The main challenge in achieving flexible teaming is coordinating the robot with operations executed by human partners. We contribute an approach to this problem that is based on observing pre- and postconditions of operations using a robot-mounted camera system. The coordination mechanism is embedded into a framework that allows domain experts to create, test, and dispatch new task models to collaborative execution. The approach is validated by experimental tasks composed of pick-and-place operations.

Keywords: human-robot collaboration, equal-partners teamwork, human-robot coordination, task modelling

1 Introduction

Traditional robot systems are designed to repeatedly perform the same task for a long period of time. The safety of human workers in industrial setups is often ensured by fences that prevent them from entering the robot workspace. Recent advances in the field of sensing enable safe human-robot coexistence without these physical barriers (e.g. by fast reconstruction of the robot workspace [26] and reactive motion planning [5] or tactile sensors [2]). We envision a hybrid assembly cell that follows the paradigm of symbiotic assembly [3] and thus uses the precision and strength of robots as well as the cognitive skills unique to humans to work together on achieving a common goal. By symbiotic combination of intelligent robotics and human knowledge and experience, systems that offer the flexibility required for applying robots in small batch production, handicraft workshops or small laboratories can be designed. We therefore propose a method that combines execution of robot operations contributing to goal achievement with operations to acquire sensor data needed for synchronizing with human actions. This mechanism is embedded into a system allowing domain experts to model and execute tasks in human-robot collaboration.

2 Related Work

We structure work related to human-robot common goal achievement based on task models according to the degree of flexibility they offer to the participating agents during task execution. Least flexibility is offered by systems that perform a **fixed assignment of operations** within a task to agents in advance of the execution process. In [10], methods from classical assembly planning are adapted to build a hierarchical framework for optimal task allocation and execution in human-robot collaborative assembly. The HRI/OS [4] uses a centralized executive to delegate tasks to humans and robots. Groups of operations can also be allocated to agents by assigning a role to each of them [13][21].

The next level in our taxonomy is formed by approaches where either a **human or robot dominates** the process, while the counterpart within the team follows orders or adapts. Control can be given to the robot, which plans, assigns and explains parts of the task to a human partner [1] [20]. The task allocation is negotiated in [7], where a robot asks the human for permission before performing an operation. Other approaches regard the robot as a tool that assists humans by performing assistive actions [6][12][16].

We regard systems with a focus on **decision authority for all agents** as the most flexible. The Chaski executive [24] enables dynamic execution of tasks formulated as Temporal Constraint Networks (TCNs) by just-in-time assignment of operations to agents. Each agent decides for the next operation on the fly based on the decisions taken and communicated by others, resulting in an execution process that fulfils all time constraints encoded in the TCN. Similar to [24], the approach of [17] uses precompiled versions of Temporal Planning Networks under Uncertainty for integrated plan recognition and dynamic execution.

Our hypothesis is that a robot system should possess the properties listed below to achieve symbiotic collaboration. These properties also allow the integration of our approach into the above taxonomy. An exchangeable *task model* should be the main input, preferably one similar to those used in assembly planning (e.g. precedence graphs, AND/OR-Graphs [8]) so that existing planning algorithms can be reused for adding knowledge to the system. The task model is *shared* between humans and robots to approximate a *shared mental model* [11]. Studies on human teaming show, that this increases team work effectiveness [18][19]. It may be necessary for human agents to handle interrupts in small workshops, e.g to serve entering clients. If the system allows for a *dynamic team setup*, meaning that the number of humans and robots can arbitrarily change, the working progress will not be stopped in such cases. This can not be achieved by a pre-computed schedule, as task allocation needs to vary depending on the current team composition. Thus, requesting flexible teams requires *dynamic plan execution*, where agents decide about their next steps on the fly [24]. Combining flexible teams and agents with strongly asymmetrical capabilities would likely stop the progress when highly skilled agents leave. We therefore focus on tasks composed of operations that can be executed by either human or robot, possibly within differing amounts of time, and refer to this as *equal partners collaboration*. Our understanding of equal partnership is based on [24], where *equal partners teamwork* means, "that each member of the team has equal authority to make decisions when executing the plan". Through this paradigm, intuition of humans about their own capabilities and preferences is utilized to achieve a workflow that is convenient for workers, e.g. regarding ergonomics. We explicitly focus small-scale scenarios, where flexibility is favored over time optimal task allocation.

As our coordination mechanism is designed to satisfy the above criteria, we sort it into the category of approaches with decision authority for all agents in our related work taxonomy. While [24] and [17] focus on fulfilling timing constraints, we explicitly enable flexible team composition. In [24], the decisions are taken based on other agents' communication, and [17] receives estimates of the world state from an external component. In contrast, our contribution plans sensor operations in addition to task operations to gather information needed for robots to select their next actions just-in-time.

3 Coordinating Flexible Human-Robot Teams

Our approach is based on *elementary operation templates* $O = \{o_1, o_2, ..., o_{|O|}\}$ that the robot is able to execute. Every operation template $o \in O$ needs parameters taken from the set P_o of valid parameter combinations to o. E.g., a pick-and-place operation may need start and goal position of an affected object. A pair $\bar{o} = (o, p)$, $p \in P_o$ is named an *operation instance*. O partitions into two subsets O_S and O_T. O_S is a set of *sensor operation templates* that are used to acquire sensor data, e.g. moving a camera and capturing an image. The *task operation templates* O_T are used to form shared task models. A *shared task model* $T = \{\bar{o}_1, \bar{o}_2, ..., \bar{o}_{|T|}\}$ is composed of several operation instances whose templates are taken from O_T. The elements of T may be part of some superordinate structure like the graph defined by precedence relations [8]. An exemplary model of a palletizing task as used in our experiments is depicted in Fig. 1.

The coordination mechanism bases upon the observation, that elementary operations usually need some *preconditions* to be satisfied before they can be executed. E.g., a pick-and-place operation requires the affected object to exist within the workspace. Preconditions we are considering are not limited to resources, as the presence of humans in the workspace might result in additional requirements. E.g., if a task includes a welding operation, robot agents need to ensure that the safety door of the welding system is closed. As soon as an operation was successful, it produces observable effects manifesting in fulfilled *postconditions*. E.g., the moved object will be at it's new position. More formally, we define the sets V_o of *pre-* and N_o of *postcondition templates* for every template $o \in O_T$. A *condition instance* $\bar{c} = (c, \bar{o})$ combines a condition template with the operation instance $\bar{o}$ it has to be evaluated for. Every operation $\bar{o}_i = (o_i, p_i)$ within a task T has a set of *preconditions* $V(\bar{o}_i)$ that contains an instance for every template in V_{o_i}. The postconditions $N(\bar{o}_i)$ are defined respectively.

A set of fulfilled pre- or postconditions allows to decide whether the corresponding operation may be started or has already been completed. Hence,

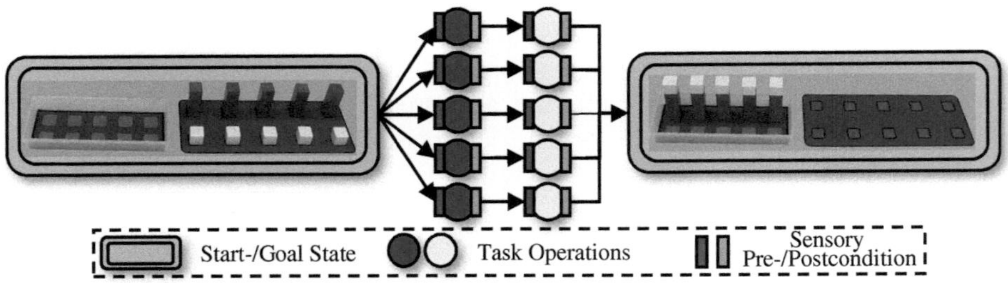

Fig. 1. Tasks can be modelled as precedence graphs that convert a start state into the desired goal state by a suitable composition of task operations. Each operation is annotated with pre- and postconditions needed for coordinating humans and robots.

perceiving the state of conditions provides a way to decide whether an operation is executable. We model the process of condition evaluation as a function $\mathcal{P}$ that abstracts the details of perception. $\mathcal{P}$ takes an arbitrary set of condition instances and returns **true**, if all of them are fulfilled, otherwise **false**. Internally, $\mathcal{P}$ may trigger sensor operations to gather suitable data for the evaluation. Given $\mathcal{P}$, *readiness* $\mathcal{R}$ and *success* $\mathcal{S}$ of an operation instance $\bar{o}_i$ are defined by

$$\mathcal{R}(\bar{o}_i) = \mathcal{P}(V(\bar{o}_i)) \qquad \mathcal{S}(\bar{o}_i) = \mathcal{P}(N(\bar{o}_i)).$$

If and only if $\mathcal{S}(\bar{o}_i) = \textbf{true}$, then $\bar{o}_i$ has already been done. Respectively, if and only if $\mathcal{R}(\bar{o}_i) = \textbf{true}$, all preconditions of $\bar{o}_i$ are fulfilled. An operation instance with $\mathcal{R}(\bar{o}) = \textbf{true}$ and $\mathcal{S}(\bar{o}) = \textbf{false}$ is named *active*. Correct task execution is guaranteed by only executing active operations.

4 Prototype System

In the following, we describe components of our prototype system based on the above coordination mechanism in detail. The main modules are depicted in Fig. 2. The workflow consists of three steps: First, the task model is created using a graphical editor (Fig. 3, left). The editor offers functionality to add task objects (blue boxes) and operations (gray boxes) working on them. Required operation parameters are input through dialog windows. A task structure is created by connecting operations using precedence graph edges. The results of task modelling can be checked by observing the robot performing the task within a simulation (Fig. 3, right). Erroneous operations or parameters can be corrected by returning to the editor. The precedence graph undergoes an automatic preprocessing step that supplements each operation with suitable pre- and postcondition instances. Then, the graph is stored as an XML file for later usage and can be passed to a module for collaborative execution. Given a specific set of sensors and actuators, this module encapsulates an algorithm to determine an efficient order of condition validation steps and execution of operations, such that the task is carried out correctly. Details of our current implementation of the execution module are described in Sections 4.1, 4.3 and 4.2.

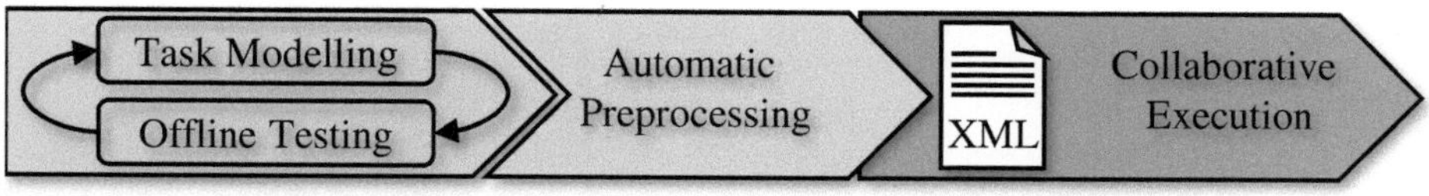

Fig. 2. The system workflow has three steps: Task models are created by repeated modelling and offline testing. A preprocessing step generates required conditions. This results in an XML task representation serving as input to the execution module.

4.1 Hardware, Supported Operations and Conditions

We use a KUKA LBR IV with a Robotiq 3-Finger gripper. Worker safety is ensured by using the robot in compliant mode and at moderate speed. Though our approach allows integrating arbitrary sensors, we focus on cameras for condition evaluation. In theory, the approach described in Section 3 could be implemented by equipping the robot cell with cameras overseeing the whole workspace at any time, and repeatedly evaluating all conditions e.g. through object recognition and localization. The set of active operations would then always be available for the system to select it's next step. We intend to manipulate small objects within a workspace, where several humans and robots might be moving. Thus, a high level of occlusion may be expected when using fixed cameras, rendering them impractical. Therefore, the robot is equipped with an IDS uEye UI-1220SE-C-HQ eye-in-hand camera, enabling it to look at specific positions for condition evaluation. Besides the problem of occlusions, this also reduces the amount of hardware within the workspace, making the installation less complex and costly.

Our set of operation templates O_T enables manipulation tasks through operations to transfer objects from fixed start to goal positions and apply a stamp to them. In this context, condition templates to evaluate, whether an object of a certain type is present at a given position or not, and whether it is stamped or not, are needed. The only required sensor operation o_S moves the robot to a given position and triggers the camera. The perception function $\mathcal{P}$ maps each input condition instance to an instance of o_S. Condition evaluation is realized through object recognition applied to the resulting image, specifically by classifying objects according to their color. Our approach can intuitively be extended to more complex scenarios by adding hardware (e.g. a tool changer), operations (e.g sensor-guided skills [25]), sensors and evaluation algorithms (e.g. active manipulation for recognition of objects [22] and material properties [28]). However, such special implementations are out of the scope of this conceptual paper.

4.2 Execution Module Software Design

The execution module is structured as a multi-agent system [9]. A *coordination agent* executes the algorithm to achieve correct task execution through task and sensor operations as described in Section 4.3. To this end, it requests evaluation of $\mathcal{R}$ and $\mathcal{S}$ from the *perception agent*, which maps the conditions to sensor operations and performs image processing. Task operations as well as sensor

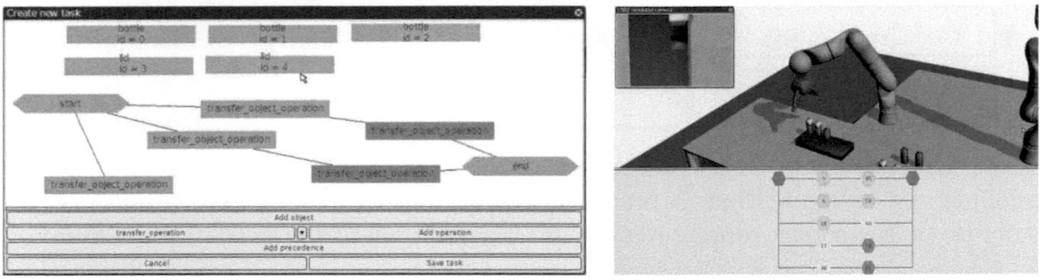

Fig. 3. The prototype system provides a graphical editor for modelling new tasks (left). The execution of tasks by a robot can be observed in a simulation environment (right) to validate correct parametrization of elementary operations.

operations are dispatched to the *hardware agent* that transforms abstract operations into concrete hardware commands.

The execution module realizes all agents through the ENACT software framework [27]. This framework follows the popular programming paradigm of minimized coupling and maximized cohesion (see [14]). To this end, ENACT offers exchangeable agents and couples these through abstract data types. This enables quick porting of the execution module to different hardware. For instance, changing the robot type is intuitively done by implementing and using another hardware agent. The same holds for new sensors and the perception agent.

4.3 Execution Algorithm for Robot-Mounted Cameras

Cameras attached to the robot can only view parts of the workspace at a time. The challenge lies in finding an order of condition evaluation and execution of operations that reduces the number of sensor operations, while still ensuring correct task execution. This can be achieved by tracking the task progress to extract operations and conditions currently relevant. We therefore integrate operations into precedence graphs (Fig. 1). The knowledge a robot has about the progress is encoded in two sets I_1 and I_2. I_1 denotes operations whose predecessors in the graph have successfully been executed. Thus, elements of I_1 are candidates for testing whether they are active. I_1 is initialized with all operations directly connected to the start node. Operations with preconditions that have been evaluated negatively and consequently are assumed to be carried out by another agent at the moment are held in I_2. Initially, I_2 is empty. The execution algorithm is visualized in Fig. 4. As long as I_1 or I_2 contain any elements, the task is not completed and the algorithm keeps running. In the *Execution Phase* (dark grey), an element $\bar{o}$ from I_1 is selected first. Currently, the selection is biased towards following branches within the precedence graph, meaning that completing subtasks is preferred. Further concepts like a capability index [23] may be integrated to e.g. prefer operations with a high automation potential. If the postconditions of $\bar{o}$ are fulfilled ($\mathcal{S}(\bar{o}) = \textbf{true}$), all successors of $\bar{o}$ whose other predecessors already have been completed can be added to I_1. Otherwise, preconditions are checked by determining $\mathcal{R}(\bar{o})$. If they are fulfilled, $\bar{o}$ is dispatched to the execution agent.

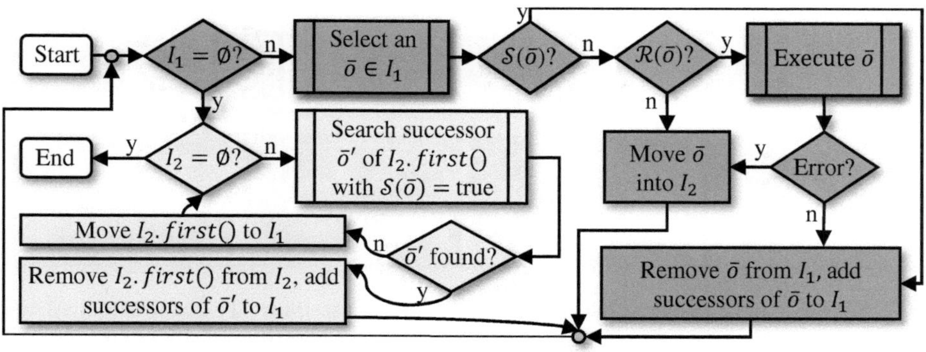

Fig. 4. Our algorithm for human-robot collaboration has two phases. In the *Execution Phase* (dark grey), the system looks for active operations and executes them. If operations with unfulfilled preconditions are detect, they are stored for further processing in the *Monitoring Phase* (light grey) that tries to detect progress and may move operations back to the Execution Phase for a retry.

We currently assume, that no errors occur during execution. This assumption can be relaxed using robust operation implementations that support rollback on errors [15]. If preconditions are not fulfilled, $\bar{o}$ is assumed to be in progress and moved to I_2 for later clarification. The Monitoring Phase is entered as soon as there are no tasks left in I_1 - the system has to synchronize with operations executed by humans during the Execution Phase. To this end, the elements of I_2 are iterated by repeatedly processing the first element $I_2.first()$ of I_2. If the postconditions of an $\bar{o} \in I_2$ are fulfilled, the successors of $\bar{o}$ are up next and can be moved into I_1 for the next Execution Phase. In addition, postconditions of the successors of $\bar{o}$ are checked. This enables the system to "catch up" and detect completed operations without having spotted their unfulfilled preconditions in the Execution Phase. This initial realization is based on the assumption, that humans participating in the collaboration are cooperative and will carry out the task correctly due to the task model. For pick-and-place operations, this means that an object that is not detected at it's start position will reappear there or will be moved to it's target eventually.

5 Experimental Validation

Exemplary tasks as used in our experimental validation are shown in Fig. 5. The experiments are targeting the evaluation of the equal-partners collaboration mode using our coordination mechanism. Therefore, we are using simply shaped, colored objects to abstract from the problems of grasp planning and object recognition. We used the GUI (Fig. 3) to create and test task models simulating assemblies (1), palletizing tasks (2) and processes involving tools like a stamp by combining the pick-and-place operations described in Section 4.1. Task 2 emulates palletizing bottles (red blocks) and sealing them with lids (white blocks). The complete model of Task 2 is shown in Fig. 1. Some steps during

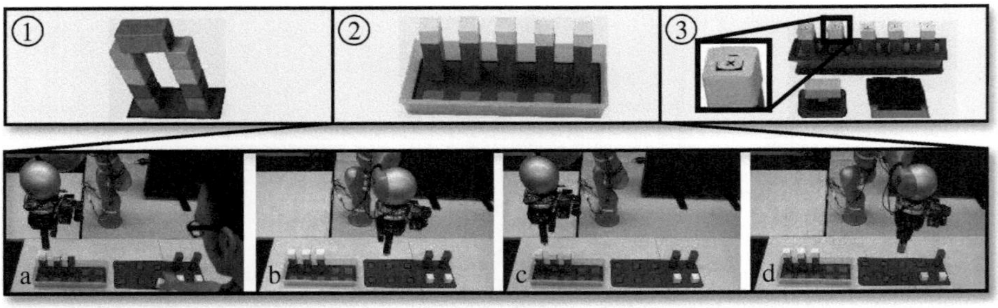

Fig. 5. Our experiments consist of setups simulating assembly (1), scalable palletizing tasks (2) and processes involving tools in addition to operating resources (3). Crucial steps during an execution of Task 2, showing how the robot detects human interaction and proceeds working, are depicted in the lower part of the figure.

an execution process are shown in the lower row of Fig. 5: The human picks up a lid, while the robot negatively evaluates the postcondition for the pick-and-place operation afflicting the same object (a). The robot moves on to check the precondition of this operation (b). As it is not fulfilled, the execution algorithm continues with another operation contained in the list of elements I_1 that are up next according to their precedence relations. In the sample case, the robot looks for the next bottle at it's target position (c). As it is not present, the precondition is checked moving to the start position and trying to locate it there. This evaluation step is successful, allowing the robot to perform the operation (d).

6 Conclusion and Future Work

We presented an approach to equal-partners human-robot collaboration that enables coordinating teams of flexible composition. The approach is based on evaluating pre- and postconditions of operations to detect human actions. It combines dynamic execution of robot operations contributing to goal achievement with operations to acquire sensor data. We integrated it into a framework enabling domain experts to create and test shared task models using a Graphical User Interface. This system can be adapted to tasks composed of arbitrary elementary operations by adding suitable sensors, actors and condition evaluation algorithms. The feasibility of the approach is shown by experimental execution of pick-and-place tasks. The tasks are formulated as precedence graphs and may be carried out by a team of humans and a robot with a camera attached.

Currently, the implementation does not make use of the fact, that one camera image might provide information about the state of several conditions. This results in unnecessary robot motion, as the system does not remember the previously seen and performs a sensor operation for every evaluation request. However, it does not suffice to extract and store maximum information from one image. Due to unpredictable human actions, the state of conditions may change with time. In our future work, we will integrate a world model emulating the process of remembering and forgetting to deal with this problem of data aging.

References

1. M. Foster and C. Matheson, Following Assembly Plans in Cooperative, Task-Based Human-Robot Dialogue, Proc. of the 12th Workshop on the Semantics and Pragmatics of Dialogue, 2008
2. M. Fritzsche, N. Elkmann and E. Schulenburg, Tactile Sensing: A Key Technology for Safe Physical Human Robot Interaction, 6th ACM/IEEE International Conference on Human-Robot Interaction (HRI), Lausanne, 2011
3. P. Ferreira, S. Doltsinis and N. Lohse, Symbiotic Assembly Systems – A New Paradigm, Procedia CIRP, vol. 17, pp. 26–31, 2014
4. T. Fong et al., The Human-Robot Interaction Operating System, Proc. of the 1st ACM SIGCHI/SIGART Conference on Human-Robot Interaction, 2006
5. T. Gecks and D. Henrich, Sensor-Based Online Planning of Time-Optimized Paths in Dynamic Environments, GWR09 German Workshop on Robotics, Germany, 2009
6. T. Hamabe, H. Goto and J. Miura, A Programming by Demonstration System for Human-Robot Collaborative Assembly Tasks, IEEE International Conference on Robotics and Biomimetics (ROBIO), Zhuhai, 2015
7. G. Hoffman and C. Breazeal, Collaboration in Human-Robot Teams, Proc. of the AIAA 1st Intelligent Systems Technical Conference, Chicago, 2004
8. L.S. Homem de Mello and A.C. Sanderson, Representations of Mechanical Assembly Sequences, IEEE Transactions on Robotics and Automation, vol. 7, no.2, 1991
9. N. Jennings, On Agent-Based Software Engineering, Artififial Intelligence, Volume 117, Issue 2, 2000
10. L. Johannsmeier and S. Haddadin, A Hierarchical Human-Robot Interaction-Planning Framework for Task Allocation in Collaborative Industrial Assembly Processes, IEEE Robotics and Automation Letters 2.1: 41-48, 2017
11. C. M. Jonker, M. B. van Riemsdijk and B. Vermeulen, Shared Mental Models: A Conceptual Analysis, COIN 2010 International Workshop, Lecture Notes in Artificial Intelligence Volume 6541, 2010
12. H. Kimura, T. Horiuchi and K. Ikeuchi, Task-Model Based Human Robot Cooperation Using Vision, Proc. of the IEEE/RSJ International Conference on Intelligent Robots and Systems (IROS), Kyongju, 1999
13. S. Lalle et al., Towards a Platform-Independent Cooperative Human Robot Interaction System: III An Architecture for Learning and Executing Actions and Shared Plans, IEEE Transactions on Autonomous Mental Development, vol. 4, no. 3, 2012
14. M. E. Latoschik and H. Tramberend: A Scala-Based Actor-Entity Architecture for Intelligent Interactive Simulations, 5th Workshop on Software Engineering and Architectures for Realtime Interactive Systems (SEARIS), 2012.
15. J. Laursen, U.Schultz and L. Ellekilde, Automatic Error Recovery in Robot Assembly Operations using Reverse Execution, IEEE/RSJ International Conference on Intelligent Robots and Systems (IROS), Hamburg, 2015
16. C. Lenz et al., Joint-Action for Humans and Industrial Robots for Assembly Tasks, 17th IEEE International Symposium on Robot and Human Interactive Communication, Munich, 2008
17. S. J. Levine and B. C. Williams, Concurrent Plan Recognition and Execution for Human-Robot Teams, 24th International Conference on Automated Planning and Scheduling (ICAPS), 2014
18. B. Lim and K. Klein, Team Mental Models and Team Performance: A Field Study of the Effects of Team Mental Model Similarity and Accuracy, Journal of Organizational Behavior, 27(4):403, 2006

19. E. Mathieu et al., The Influence of Shared Mental Models on Team Process and Performance, The Journal of Applied Psychology, 2000
20. G. Milliez et al., Using Human Knowledge Awareness to adapt Collaborative Plan Generation, Explanation and Monitoring, 11th ACM/IEEE International Conference on Human-Robot Interaction (HRI), Christchurch, 2016
21. S. Nikolaidis and J. Shah, Human-Robot Cross-Training: Computational Formulation, Modeling and Evaluation of a Human Team Training Strategy, 8th ACM/IEEE International Conference on Human-Robot Interaction (HRI), Tokyo, 2013
22. A. Schneider, J. Sturm et al., Object Identification with Tactile Sensors using Bag-of-Features, IEEE/RSJ International Conference on Intelligent Robots and Systems (IROS), St. Louis, 2009
23. D. Schroeter et al., Methodology to Identify Applications for Collaborative Robots in Powertrain Assembly, Procedia CIRP, vol. 55, pp. 12–17, 2016
24. J. Shah et al., Improved Human-Robot Team Performance using Chaski, a Human-Inspired Plan Execution System, Proc. of the 6th International Conference on Human-Robot Interaction (HRI), Lausanne, 2011
25. U. Thomas, B. Finkemeyer, T. Kroger and F. M. Wahl, Error-Tolerant Execution of Complex Robot Tasks based on Skill Primitives, IEEE International Conference on Robotics and Automation (ICRA), Taipei, 2003
26. T. Werner and D. Henrich, Efficient and Precise Multi-Camera Reconstruction, 8th ACM/IEEE International Conference on Distributed Smart Cameras (ICDSC), Venice, 2014
27. T. Werner et al., ENACT: An Efficient and Extensible Entity-Actor Framework for Modular Robotics Software Components, 47th International Symposium on Robotics (ISR), 2016
28. H. Yussof et al., Low Force Control Scheme for Object Hardness Distinction in Robot Manipulation based on Tactile Sensing, IEEE International Conference on Robotics and Automation (ICRA), Pasadena, 2008

Virtual Testbeds for the Development of Sensor-Enabled Applications

Jörn Thieling[1] and Jürgen Roßmann[2]

Institute for Man-Machine-Interaction
RWTH Aachen University
Ahornstrasse 55, 52074 Aachen, Germany

[1]thieling@mmi.rwth-aachen.de,[2]rossmann@mmi.rwth-aachen.de

Abstract. The development of new technical systems is a cost- and time-intensive process. For this reason, Digital Prototyping and simulation technologies are widely used to enable efficient system tests in all stages of development. At MMI, we propose the use of Virtual Testbeds (VTBs), in order to allow for the verification, validation and optimization of the entire system. VTBs are an important tool to model, test and optimize a technical system and its environment in a comprehensive 3D simulation, before building a real prototype. In our contribution, we describe the integration of various sensor simulation aspects into a VTB in terms of different sensor types, analysis tools, data processing pipelines and visualization opportunities. This includes a framework that supports not only simulated but also real sensors as well as a design that connects various libraries to maximize the opportunities of sensor data processing and analysis within a VTB.

Keywords: Virtual Testbeds, Sensor Systems, 3D Simulation, HiL, SiL

1 Introduction

Today, Digital Prototyping and simulation technologies are common tools in the development process of new technical systems. They allow for the verification and optimization of system components and algorithms in all stages of the development even before a real prototype has been build. Thus, the development time can be shortened, the costs can be reduced and the product quality can be improved. In order to get a holistic view on the technical system and its behaviour in a specific environment we propose to model and simulate the entire system and its prospective environment instead of simulating the components on their own. Thus, this paper focuses on the new concept of Virtual Testbeds for sensor-enabled applications. Virtual Testbeds (VTBs) are comprehensive 3D simulation environments which reallistically model all important aspects of a given application [1]. This comprises a detailed simulation of the system itself (e.g. its dynamics, sensors and actuators) as well as an adequate simulation of the environment including all interactions between system and environment.

Using Virtual Testbeds for the development of sensor-enabled applications allows for a flexible way of testing the system design in any multi-domain environment followed by a target-oriented optimization of the entire system. This concerns the selection and configuration of hardware components (e.g. sensors and actuators) as well as the algorithms of the data processing system. Besides, VTBs support hybrid system tests by combining real and virtual components as available or required. This successive transition enables the direct use of developed algorithms and models for operation. For instance, the mobile robot in figure 1 uses a VTB and the data-processing pipeline previously developed in the VTB by feeding it now with real instead of simulated sensor data. Each part of this system including dynamics, sensors, actuators and data processing are first modelled and simulated in the VTB and then successively replaced by their real counterpart.

Fig. 1. Real laser scanner mounted on a mobile robot driving through a factory (left). Visualization of the resulting sensor data after an automated alignment process (right).

In this contribution, we will start with a short overview of the state of the art in simulation of sensor-enabled applications followed by the key concepts of VTBs and their comprehensive Sensor Framework, which allows for a modular and flexible way of sensor data processing and the assisted generation of validated sensor models. The contribution closes with an overview of selected applications and a short conclusion.

2 State of the Art

Taking a look at today's simulation technologies reveals a broad range of different tools and approaches. Starting with very general approaches such as MATLAB/Simulink [2] or Modelica/Dymola [3] which allow for block-oriented simulation of multi-domain systems, advanced simulation tools rapidly become application-specific. For industrial applications, Plant Simulation [4] and FlexSim [5] are examples, which provide process and logistic simulations but completely ignore the sensor aspect of modern plants. But even if focusing on the robotics part where sensors are indispensable in times of intense automation and autonomous robots, very well-known products like Microsoft Robotics Studio [6],

Gazebo [7], Webots [8] and USARSim [9] rarely allow for a comprehensive and validated sensor simulation. A look at the automotive sector reveals various tools which focus on testing of control software as well as tools which aim at future validation of advanced driver assistance systems, examples are dSPACE [10], CarMaker [11], PreScan [12] and Pro-SiVIC [13].

The drawback of nearly all of these systems is the loss of flexibility. Either they focus on simulation of multi-domain applications but do not allow for advanced sensor simulation or they provide simulation of sensor-enabled systems but only in a very specific field of application. This leads to the problem that the development of systems applicable to multiple domains (e.g. a sensor designed for the use in urban scenarios as well as factory buildings) requires multiple simulation tools resulting in inconsistent system models caused by different levels of detail. Moreover, the capabilities of generating validated sensor models by means of sensor calibration are extremly limited. Often, there is a huge gap between the virtual and real technical system and its environment instead of a seamless transitions for a continuous transfer between digital and real prototype.

3 Virtual Testbed Concept

As mentioned in the introduction, Virtual Testbeds realistically model all important aspects of a given application. Therefore, the desired technical system can be modeled as a configuration of simulated sensors, actuators and further components (e.g. mechanical parts) represented by their state vectors $\underline{s}\,^{sim}_{sense}(t)$, $\underline{s}\,^{sim}_{act}(t)$ and $\underline{s}\,^{sim}_{sys}(t)$ – as depicted in figure 2.

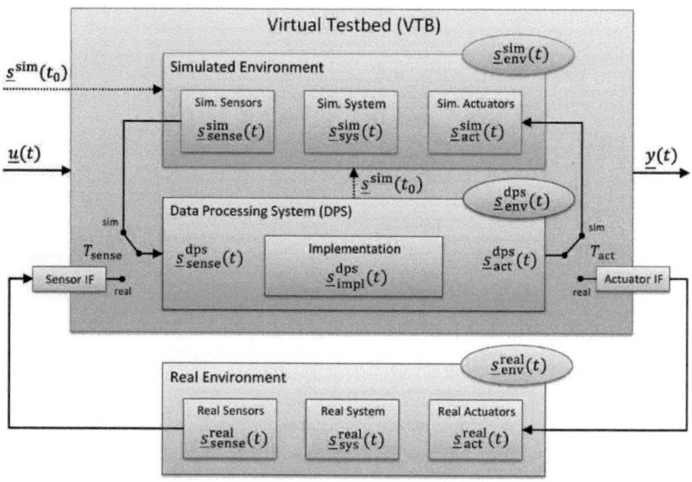

Fig. 2. Virtual Testbed Structure. [1]

The resulting model of hardware components can then be placed in any simulated environment $\underline{s}^{sim}_{env}(t)$ while the corresponding software parts of the technical system are combined in a separate Data Processing System (DPS) represented by its implementation $\underline{s}^{dps}_{impl}(t)$, sensor input $\underline{s}^{dps}_{sense}(t)$, actuator output $\underline{s}^{dps}_{act}(t)$ and perceived environment $\underline{s}^{dps}_{env}(t)$. Due to the comprehensive set of interfaces for real sensors and actuators, the Data Processing System can be used with simulated as well as real technical systems by turning the switches T_{sense} and T_{act}. This is a key idea of Virtual Testbeds to allow for a seamless transition between simulation and reality in all stages of development. Thus, data processing pipelines can be initially designed, tested and optimized with the simulated system and afterwards directly used for controlling the real prototype [14].

4 Sensor Framework

Due to the fact that most of today's technical systems are using sensors to perceive their environment or measure their internal states, sensor simulation and data processing is an important aspect for Virtual Testbeds. Thus, a comprehensive Sensor Framework [15] has been developed to meet the needs of nearly all kinds of sensor-enabled applications independent of their domain.

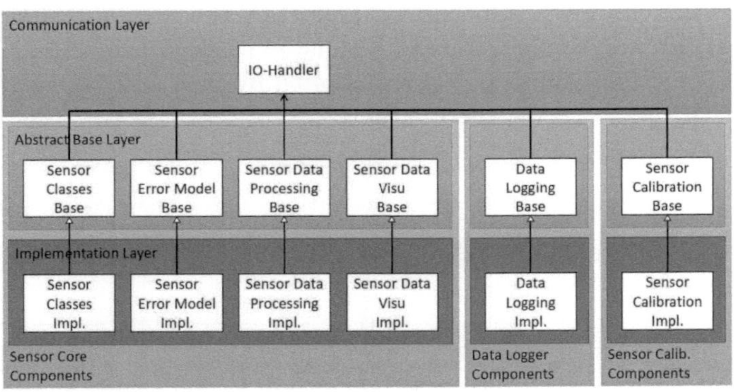

Fig. 3. Sensor Framework Architecture (based on [15]).

Figure 3 gives an overview of the Sensor Framework Architecture which has been completely integrated in the aformentioned Virtual Testbed Concept. The architecture consists of three different layers. The first one is the communication layer which provides an unconfined way of interaction between all modules within the Virtual Testbed. Therefore, generic IO-Boards with standardized interfaces such as sensor data inputs and outputs are used. The second layer comprises the abstract base classes for all kinds of Sensor Framework Components while their specific implementations are placed in the third layer. This ensures that basic inputs and outputs required for all inherited components are created in a standardized way. For instance, all supported 3D laser scanners inherit from the same base class. Hence, they have the same interface no matter if they are simulated or real. In order to allow for a comprehensive simulation, verification and optimization of sensor-enabled applications the Sensor Framework comprises the following key components (see Fig. 3):

Sensors of various types such as cameras, laser scanners, ultrasonic sensors, GPS receivers, IMUs, radar sensors and much more are building the base by generating a broad range of sensor data. For each of these sensor types several variants of simulated (varying level of detail) as well as real sensors (varying manufacturer) are provided while the aformentioned standardized interface is building their focal point. While the real sensors are connected via their device-specific interface (e.g. UDP socket) the virtual sensors mostly are simulated by using the GPU to accelerate their computation without loss of precision.

Error Models for each sensor type extend the capabilities of simulated sensors by adding parametrizable inaccuracy and typical effects to their initially ideal sensor data. This allows for close-to-reality sensor models which adequately mimic their corresponding real devices (see Fig. 4). For instance, a simulated camera can be extended by various error models such as lens distortion, noise, vignetting, depth of field, motion blur and much more [16]. Besides, all of these camera effects are computed on the GPU to enable real-time simulations.

Fig. 4. Smear effect of real and simulated camera capturing a light source. [16]

Sensor Data Processing is part of every sensor-enabled application. Thus, the Sensor Framework provides a huge set of sensor data processing components (e.g. sensor data filters and analysis tools) which can be used for a fast and flexible development of powerful data processing pipelines within the Virtual Testbed. For this purpose, various third party libraries from multiple domains have been linked to our Sensor Framework. For instance, the Open Source Computer Vision library (OpenCV) [17] which provides several state-of-the-art algorithms for image processing or the Point Cloud Library (PCL) [18] which provides various point cloud processing algorithms. All of these algorithms are not only provided as pure functions which can be used to program further advanced function blocks but also as ready-to-use data processing components which can be used for a modular and flexible generation of any desired data processing pipeline. For instance, a data processing pipeline consisting of minimum intensity filter, point cloud converter, outlier removal, smoothing and triangulation can be modelled by connecting and parametrizing the corresponding components (see Fig. 5).

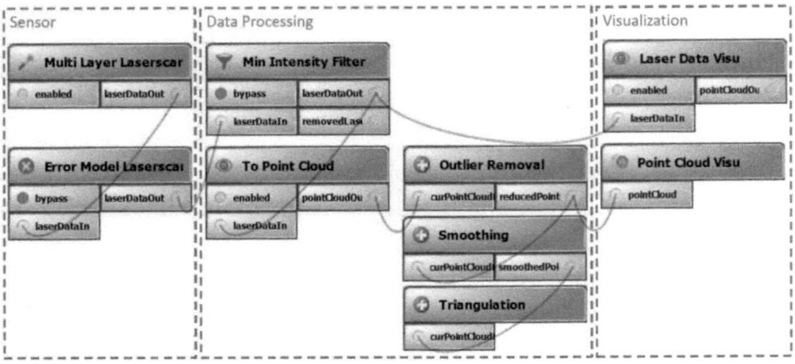

Fig. 5. IO-Editor's view on an exemplary system consisting of laser scanner, appropriate error model, data processing pipeline and visualization modules within a VTB.

Sensor Data Visualization is another very important aspect in the verification and optimization process of sensor-enabled applications. Usually, modern applications handle a huge amount of sensor data from several sensor devices which is hardly comprehensible without further tools. Thus, our Sensor Framework provides various visualization opportunities for all kinds of sensor data types to allow for an intuitive way of sensor data analysis. For instance, laser data can be visualized as fading rays as well as color-coded point clouds by using real-time rendering techniques (see Fig. 7a).

Data Logging is essential in terms of saving simulation results or recording real sensor data for later analysis. Thus, additional logging components allow for writing every kind of data from the Virtual Testbed to a desired logfile using the standardized interfaces introduced above. Besides, these components allow for the playback of complex sensor simulation results in real-time. For instance, a highly detailed simulation of multiple sensors and complex systems which hardly runs in real-time can be simulated overnight and afterwards visualized and analyzed in a real-time 3D virtual environment.

Sensor Calibration is indispensable to achieve a reliable sensor simulation, but at the same time it is very time-consuming and error-prone. Often, slight differences in the measurement process lead to notable deviations in the calibration results. For instance, camera calibration comprises taking different pictures of a chessboard while varying the relative pose between camera and chessboard. As a consequence, consecutive calibrations of the same camera usually lead to different picture sequences used for the calibration algorithm which results in inconsistent sensor models. In order to improve this process of transfering sensor devices between reality and simulation, our Sensor Framework includes not only components providing calibration algorithms but also components assisting the user in generating validated sensor models. For instance, components for camera calibration automatically generate appropriate poses for camera and chessboard taking into account the already known properties (e.g. chessboard size and camera field of view). Afterwards another component controls the actuators of a physical mockup, such as the one depicted in figure 6, to relocate camera and chessboard according to the previously calculated poses. Finally, the calibration algorithm component calculates the lens distortion parameters which can directly be used for setting up the corresponding error model to generate validated sensor models in a comfortable, fast and reproducable manner.

Fig. 6. Physical mockup used for the assisted generation of validated sensor models.

5 Applications

The Virtual Testbed approach introduced above has been already used for various sensor-enabled systems from different fields of application (see Fig. 7). Currently, these applications are mainly settled in one of the major sectors: space, automotive and industry. The following lines will give a glance at each of them.

Space: Space missions are very cost-intensive and test scenarios comprising zero gravity and space environments are hardly realizable on earth. Here, the automated docking scenario of the ATV at the ISS can be efficiently tested, verified and improved by simulating different system designs and approach scenarios within a VTB including the interactions between space dynamics and sensor measurements (see Fig. 7a).

Automotive: Currently, sensor data for millions of driving kilometres in urban scenarios are recorded for the validation and optimization of advanced driver assistance systems. The drawback of these real test drives is their huge time-consumption and missing focus on rare but very important critical situations (e.g. pedestrians crossing street at red light). Here, VTBs can be used for the repeatable and fully automated generation of sensor data for various different scenarios which can be either explicitly specified or randomized (see Fig. 7b).

Industry: Due to the massive automation of modern plants, the development and optimization of sensor-enabled industrial systems becomes increasingly important. Here, VTBs can be used for determining optimal placement of sensors within a system taking into account the effect of robot and sensor poses on available sensor data (see Fig. 7c). Moreover, driverless transport systems can be tested and optimized without damaging material or persons (see Fig. 7d).

(a) Simulated 3D laser scanner mounted on an ATV.

(b) Simulated stereo camera and multi layer laser scanner mounted on a vehicle.

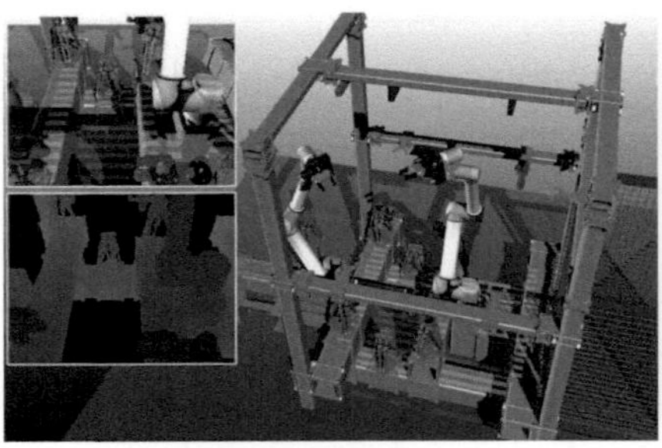

(c) Simulated PMD-sensor and industrial camera mounted on a robot cell.

(d) Simulated 2D laser scanners mounted on a driverless transport system.

Fig. 7. Sensor-enabled applications realized so far by using VTBs.

6 Conclusion

The applications presented above clearly show the flexibility and universality of Virtual Testbeds for the development of sensor-enabled applications. VTBs greatly enhance all stages of the development process by providing close-to-reality simulations of complex sensor-enabled systems in multi-domain scenarios which allow for the efficient verification, validation and optimization of entire systems taking into account all interactions with their environment. Moreover, VTBs are bridging the gap between simulation and reality by easing the transfer of technology between real and digital prototype and providing a powerful tool for flexible modeling and analysis of complex data processing systems.

7 Acknowledgments

This work is part of the project "iBOSS-3", supported by the German Aerospace Center (DLR) with funds of the German Federal Ministry of Economics and Technology (BMWi), support code 50 RA 1504.

References

1. J. Rossmann, M. Schluse, M. Rast, and L. Atorf, "eRobotics combining electronic media and simulation technology to develop (not only) robotics applications," in *E-Systems for the 21st Century – Concept, Developments, and Applications* (S. Kadry and A. El Hami, eds.), vol. 2, ch. 10, Apple Academic Press, Dec 2016. ISBN: 978-1-77188-255-2.
2. MathWorks, "Simulink: Simulation and model-based design." URL: https://mathworks.com/products/simulink.html, last visited: 13.12.2016.
3. P. Fritzson, *Principles of Object-Oriented Modeling and Simulation with Modelica 2.1.* Wiley-IEEE Press, 2004. ISBN: 978-0471471639.
4. Siemens, "Tecnomatix plant simulation," URL: https://siemens.com/tecnomatix, last visited: 13.12.2016.
5. W. B. Nordgren, "Flexsim simulation environment," in *Proceedings of the Winter Simulation Conference*, vol. 1, pp. 250–252, 2002.
6. J. Jackson, "Microsoft robotics studio: A technical introduction," *IEEE Robotics Automation Magazine*, 2007.
7. N. Koenig and A. Howard, "Design and use paradigms for gazebo, an open-source multi-robot simulator," in *2004 IEEE/RSJ International Conference on Intelligent Robots and Systems (IROS) (IEEE Cat. No.04CH37566)*, pp. 2149–2154 vol.3, 2004.
8. O. Michel, "Webots: Professional mobile robot simulation," *Journal of Advanced Robotics Systems*, vol. 1, no. 1, 2004.
9. S. Carpin, M. Lewis, J. Wang, S. Balakirsky, and C. Scrapper, "Usarsim: a robot simulator for research and education," in *Proceedings 2007 IEEE International Conference on Robotics and Automation*, 2007.
10. dSPACE, "Automotive simulation models (asm): Real-time models for vehicle development." URL: https://dspace.com/asm, last visited: 13.12.2016.
11. IPG-Automotive, "Carmaker." URL: https://ipg-automotive.com/products-services/simulation-software/carmaker/, last visited: 13.12.2016.
12. R. Bours and M. Tideman, "Simulation tools for integrated safety design," 2010. URL: https://www.tassinternational.com/prescan-press, last visited: 13.12.2016.
13. D. Gruyer, M. Grapinet, and P. D. Souza, "Modeling and validation of a new generic virtual optical sensor for adas prototyping," in *2012 IEEE Intelligent Vehicles Symposium*, pp. 969–974, 2012.
14. T. Cichon, M. Priggemeyer, and J. Rossmann, "Simulation-based control and simulation-based support in erobotics applications," in *Proceedings of the 1st MHI Fachkolloquium der Wissenschaftlichen Gesellschaft für Montage, Handhabung und Industrierobotik 2016 (MHI)*, 2016.
15. M. Emde, M. Priggemeyer, T. Steil, G. Grinshpun, and J. Rossmann, "A virtual space robotics testbed for optical sensors in aerospace," in *Proceedings of the 47th International Symposium on Robotics - Robotics in the era of digitalisation (ISR 2016)*, 2016.

16. T. Steil and J. Rossmann, "A virtual reality testbed for camera simulation in aerospace applications," in *Proceedings of the IEEE Third International Conference on Artificial Intelligence, Modelling and Simulation (AIMS 2015)*, IEEE, 2015.
17. G. Bradski, "The opencv library," *Dr. Dobb's Journal of Software Tools*, 2000.
18. R. B. Rusu and S. Cousins, "3d is here: Point cloud library (pcl)," in *IEEE International Conference on Robotics and Automation (ICRA)*, (Shanghai, China), May 9-13 2011.

Cloud-based Plant and Process Monitoring based on a Modular and Scalable Data Analytics Infrastructure

Ilja Maurer, Mauro Riva, Christian Hansen, and Tobias Ortmaier

Gottfried Wilhelm Leibniz Universität Hannover, Institute of Mechatronic Systems
Appelstraße 11a, 30167 Hannover
ilja.maurer@imes.uni-hannover.de,
WWW home page: https://www.imes.uni-hannover.de/

Abstract. In order to meet recent challenges for more efficient and economic industrial manufacturing plants and processes, new and already existing infrastructure undergoes transformations towards so called 'Smart Factories'. In this paper a fully integrated Data Analytics Infrastructure is introduced, which is applicable for different use-cases. The modular and scalable infrastructure basically consists of embedded devices for the acquisition of controller signals and process data from the real-time field bus, and a 'private cloud' server with high storage and computing capacity for data administration, analytics and various other services. The infrastructure's potential is demonstrated by an exemplary use-case, an energy management approach for multi manipulator handling processes, including monitoring and process optimization functionalities.

Keywords: Industrial Automation, Smart Factory, Cloud-based Data Analytics, Process Monitoring, Energy Management.

1 Introduction

The integration of digitalization in manufacturing, production and logistics transforms traditional factories into 'Smart Factories', with significant economic potential. These context-sensitive and decentralized production systems use network structures to exchange information between plants and their components and enable superordinate data analysis to provide useful insights into internal states, processes and plant's condition. This increased transparency gained through the interconnection of machines, plants and factories enables condition monitoring, product quality prediction and data-driven information extraction for dynamic re-engineering of processes and fast reaction to disruptions or failures [1]. The improved performance can reduce costs, optimize resource planning and enhance product quality [2].

Some of the main challenges towards the Smart Factory result from diverse communication standards, proprietary information and automation systems, heterogeneous data structures and interfaces [3] as well as proper approaches and tools to turn huge amounts of data into useful information [4].

This paper presents a universal and scalable Data Analytics Infrastructure, designed to cope with these challenges and meet the needs of factories towards digitalization and so called Industrie 4.0. The proposed architecture has the ability to seamlessly ingest and synchronize data from different sources (e. g. programmable logic controllers (PLCs), actuators and sensors), extract information and display analysis results. The infrastructure is designed to handle huge amounts of data and host data-driven analytics applications, representing enablers of Smart Factories.

The article is organized as follows. Section 2 provides a detailed view on the technical challenges towards Industrie 4.0. The developed infrastructure to overcome these challenges is introduced in section 3. An exemplary data analysis application in form of an energy management interface for an experimental laboratory packaging line is proposed in section 4. Section 5 closes the article with a conclusion and prospects for future work.

2 Challenges of Smart Factories

Modern industrial processes are characterized by an increase of automation and communication. Arising challenges can for instance be divided into 'three **A**s of digitalization': First, the data **Acquisition** from various sources has to be managed. Second, the permanently growing amount of variable data requires to be handled efficiently by data **Administration**. Third, information extraction has to be achieved by appropriate data-driven **Analysis** methods.

2.1 Data Acquisition

Networking across numerous factory components and the preferably seamless acquisition of process, environment, control, material and product data is one major requirement of digitalization. The widely used TCP/IP protocol has lacks of real-time response. Therefore, various fieldbus protocols are used to achieve real-time performance, e. g. EtherCAT [5]. These Ethernet-based industrial protocols each possess both strengths and weaknesses, so that combinations of different variants are often used in companies to meet all needs [6]. One important challenge of factory digitization is the integration of various communication standards to enable horizontal (e. g. machine to machine) and vertical (e. g. machine to enterprise resource planning systems) interconnectivity. So far, no unified approach exists, that allows data acquisition across heterogeneous networks for universal applications, covering real-time functionality [7]. Therefore, modern control system protocols, e. g. OPC UA unify the connection to the production process and establish as the de facto communication standard for the Smart Factory (e. g. MoMo - MobileMonitor, [8]). Nevertheless, companies that put efforts into system and automation networks may be reluctant to replace technologies that received significant investment [3]. Firmware upgrades of PLCs to support new communication protocols might not always be possible. Here, retrofitting may be achieved by the integration of embedded devices without modification of existing hardware and software.

2.2 Data Administration

Handling large amounts of data is known as Big Data [4]. There is no clear definition of Big Data, but the most common IT characterization compared to traditional data is the 'Vs' model. According to this scheme, Big Data consists of three components: volume, variety and velocity [9].

Volume refers to the massive amount of all types of data, generated from various sources. **Variety** follows from the diverse sources data is gathered from, e. g. sensors, industrial automation and control systems [10]. The collected information will be of heterogeneous nature both structured (e. g. dates, numbers and words in a fixed format) and unstructured (e. g. videos or data that do not follow a specific format) [9, 11]. Therefore, raw data will rarely be ready for analysis and has to be preprocessed first. **Velocity** corresponds to the ability of a system to satisfy strict timing requirements in the range of milliseconds (up to 1kHz and below) [12]. Later, two other **V**s were added to the model: **Value**, referring to the extraction of hidden information, knowledge and gaining of benefits from data [13] and **Veracity**, related to the importance of data quality [14]. Data Administration is not static. The increasing amount of sensors, controllers etc. requires dynamic scalability of hard- and software infrastructure across multiple computers [3]. Another huge concern is the security of the data infrastructure. Privacy and data ownership are additional challenges that need to be considered when designing a cloud-based data processing system.

In order to store, process and utilize the typically large amounts of data, powerful servers are needed to handle the Big Data challenge. These servers are often referred as cloud, representing a further core component of the infrastructure. A cloud is a decentralized system that offers distributed storage and computational resources enabling scaled, reliable and efficient data processing. In the industrial environment, a cloud unifies data standards and facilitates integration and collaboration of digital information between several machines, plants, sites and companies. Additional requirements may be the time synchronization between data sources (e. g. multiple PLC environments or fieldbuses), the hosting of data-driven analytical methods (via batch and stream modes) and simple access to analysis results preferably with low latency. Additionally, with the application of private clouds users try to manage and mitigate security risks and the question of data ownership. First approaches towards the control of robots and machine tools with an extended factory cloud are presented in [15] and [16]. This paper focuses on data-driven process modeling, visualization and derived information extraction.

2.3 Data Analysis

Connecting machines and computers only will not increase efficiency and profitability of manufacturing processes. Plant and process data can be a source of valuable information, but has to be analyzed in order to be useful. Traditional methods, based on statistics and iterative algorithms (e. g. square root algorithms) applied to small amounts of data are not applicable to the complexity of Big Data problems. Innovative tools and technologies are necessary to enable efficient computation, appropriate representation and complete understanding of the data [17] to convert it into useful information [4]. Although Data Science methods are already applied by many IT companies nowadays and are implemented successfully on human-generated or human-related data (e. g. recommendation systems), the integration into modern production industries on machine-generated data is still far from the state of the art [18].

3 Data Analytics Infrastructure

This chapter presents the main features and the enabling technologies of the proposed Data Analytics Infrastructure. The demands of the infrastructure are shown in Fig. 1 (I–IV), beginning with the seamless integration into the shopfloor to read data from industrial control system protocols (e. g. OPC UA) and fieldbuses (e. g. EtherCAT) in fieldbus-cycle-time (I). Moreover, a low cost hardware solution and data preprocessing possibilities are required, e. g. to reduce the amount of data being sent to the cloud (II). To extract valuable information from the decentralized data of various sources, prior appropriate data synchronization, storage and analysis need to be performed (III) and appropriate presented (IV). To realize this bottom to top concept a proper combination and implementation of hard- and software components is essential. The proposed Data Analytics Infrastructure includes the data collecting and preprocessing by embedded systems as Data Acquisition Units (DAUs) and the data handling and administration using a cloud-infrastructure, described below.

An important challenge of data acquisition in the industrial field is the enabling of interconnectivity. As shown in Fig. 1, at the bottom layer (shopfloor) data is produced in different units (e. g. sensors, PLCs and higher-level controllers) that communicate through various protocols and standards, based on manufacturers or initial implementations. The developed DAU seamlessly collects data from different sources and enables the integration of new industrial standards (e. g. OPC UA) and sets compatibilities to established fieldbuses (e. g. EtherCAT). The main advantage of such an external system is that minimal intervention in the controller software is necessary (slave, variable mapping) to enable data acquisition. Acting as an EtherCAT slave or OPC UA client, the DAU is able to ingest raw data from the fieldbus and PLCs without disturbing or delaying of the industrial networks. Furthermore, the deployment of this system directly at the fieldbus-layer allows for high sampling rates up to 1kHz and fast reaction to signal events. As data tends to grow in volume and speed, additional filter and compression methods are implemented on the DAU to preprocess raw data and satisfy security aspects (e. g. data anonymization and encryption). First, a data cleaning process filters redundancies and clears incompleteness of the heterogeneous data. Then, raw data is pre-analyzed using preprocessing functions (e. g. feature extraction, modeling, filter). To facilitate communication, the pre-analyzed data is converted into an open standard (JSON) and buffered for an adjustable time. Afterwards the buffered data is compressed, anonymized and finally transmitted to the cloud by a messaging system (Apache Kafka producer). The transfer of buffered preprocessed data blocks to the cloud no longer depends on fieldbus-cycle-time and thus is accomplished via TCP Sockets. Furthermore, the implemented DAU requires minimal configuration, is user-friendly and can be remotely configured via a web interface.

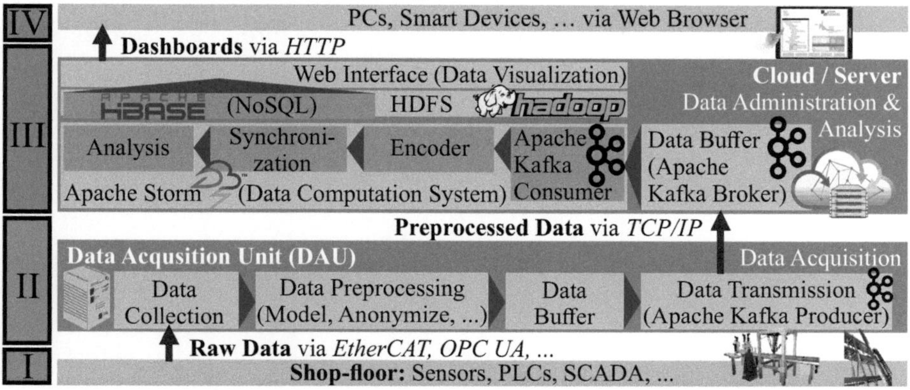

Fig. 1: Dataflow representation of the Data Analytics Infrastructure

To synchronize, store and analyze large amounts of data, appropriate tools are necessary. The most extensive and common adaption of Big Data in software is called Apache Hadoop. Apache Hadoop [19] is a java-written open-source project hosted by the Apache Software Foundation that enables the efficient data management and fast processing of large volumes of heterogeneous data across multiple computers [20]. The Apache Hadoop ecosystem contains various frameworks and services for different tasks. The main component is the Apache Hadoop Distributed File System (HDFS), designed to run on top of the local file system, splitting large files into blocks across multiple hardware [21]. To store and access recorded and analyzed data, a database system on top of HDFS is necessary. NoSQL database systems overcome the challenge of Big Data management, not requiring fixed table schemata and scaling horizontally for larger tables without performance loss [22], well integrated into the Apache Hadoop ecosystem. Although these technologies have proven to be successful in the IT sector, in industrial fields they have not been universally applied.

The developed private cloud platform enables the efficient administration and analysis of large amounts of data. Data processing and analysis inside the cloud is accomplished by the following open-source components from the Hadoop ecosystem: Apache Kafka [23], Apache Storm [24], Apache HDFS [19], Apache HBase [25] and Apache Solr [26], as described below. An Apache Kafka broker listens to the encoded messages at the cloud side and buffers them. All incoming data is decoded (deanonymized), quality checked and temporarily buffered. This approach ensures high speed data streaming and fault tolerance, meaning not to lose any data even on significant network load. Next, the messages are processed inside a computation system using Apache Storm, which enables fast handling of large data streams. Therefore, first all buffered messages are read by a Kafka consumer inside the Storm architecture. Then synchronization of the decentralized data streams of several devices is performed. Afterward the data quality is checked and cleaned when necessary. Next, use-case specific modeling, transformations and analysis methods are applied to extract value from the data. The analysis results as well as the raw data is saved in a distributed database. Due to the performance, reliability and high availability Apache HBase (NoSQL) was chosen to store the large amount of data. Finally, an appropriate representation and interaction with the stored information is necessary. Apache Solr provides a web service to query HBase for data visualization, which can be remotely accessed via PCs and smart devices. The visualization is based on a web interface, providing various possibilities to display analysis results and hence supporting effective decision-making.

4 Exemplary Use-Case: Cloud-based Energy Management

For experimental evaluations and to demonstrate the potential of the presented Data Analytics Infrastructure the components have been installed in a laboratory model factory (Fig. 2(a)), that was earlier presented in [27]. The test rig basically consists of three different robotic manipulators (4-dof SCARA, 4-dof delta robot and 3-dof small scale stacker crane) and two belt conveyors that are programmed to execute a typical handling and packing process.

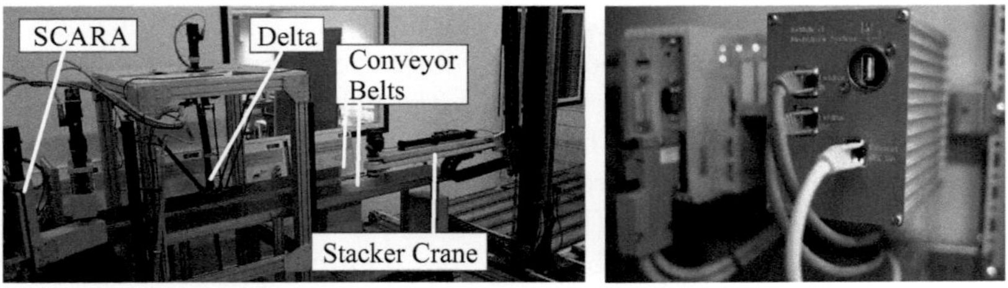

(a) Laboratory model factory (b) Data Acquisition Unit

Fig. 2: Laboratory model factory and the Data Acquisition Unit (DAU)

Using this environment, different research projects are carried out that include online system and parameter identification methods, robot trajectory planning for minimal energy and vibration reduction as well as superordinate process optimization with different objectives, e. g. minimum cycle times or energy demand reduction. Based on the Data Analytics Infrastructure (Fig. 1) different additional use-cases are under actual development, e. g. cloud-based process monitoring and energy management methods, fault tolerant control, machine learning methods for predictive maintenance of electrical servo drives and mechanical elements, etc. The following results show the benefits of superordinate plant and process monitoring and analysis of energy consumption in complex automation systems, especially in the context of alternative energy supply concepts. In this use-case no additional sensors or measuring equipment are required. The results are solely based on large amounts of heterogeneous process data obtained from the industrial controllers (e. g. PLCs and SCADA) with high sampling rates, representing a Big Data challenge (see **V**s in section 2.2).

As shown in Fig. 3, the integrated machine modules are separately controlled by industrial PLCs (module 1: delta robot and two belt conveyors, module 2: stacker crane), while the process sequence is operated and monitored by an additional higher level controller (SCADA). Despite the state of the art, all servo inverters share the same rectifier for energy supply via an interconnected DC link with energy exchange possibilities. Recuperative energy produced by a manipulator (e. g. during deceleration phases and/or motion towards lower gravitational potential) is stored in the DC link capacitance and is available for other system modules. However, if the DC link voltage reaches a threshold, the brake resistor is activated and excessive energy is dissipated. This irreversible energy loss is neither efficient nor wanted.

To setup the data **Acquisition** two DAUs are connected as EtherCAT gateways to the PLCs of both modules respectively to obtain the real-time process data of the servo drives (motor speeds and torques) and to calculate the mechanical energy conversion of each manipulator axis. Based on information from data sheets only, simple physical models [28] of the motors and inverters are implemented to calculate the inverter power in- and outputs considering energy losses of the servo drives. For this, the data preprocessing computing capacity of the DAU is utilized to analyze the continuous data stream in fieldbus-cycle-time and, afterwards, is send to the private cloud via TCP/IP in appropriate data packages to minimize data overheads. An additional DAU (Fig. 2(b)) is connected to the SCADA via OPC UA in order to capture (measure) the supply power of the plant for model validation purposes only.

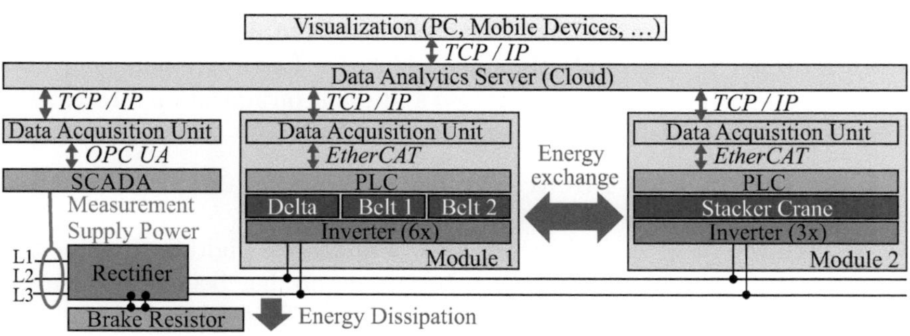

Fig. 3: Overview of energy supply concept and communication setup

For data **Administration** and the energy monitoring functionality on the cloud server level, data received from multiple decentral DAUs is time-synchronized to calculate the total DC link power and energy consumption per motion cycle of the handling process. A physical model of the DC link enables to compute different system states that are essential to evaluate the current total energy consumption and other conditions, e. g. energy dissipation, energy exchange, energy losses etc. which are usually not (cost-effectively) measurable. An interface to display the energy conditions is accessible via internet. Fig. 4 presents a sample dashboard containing two different process sequence scenarios, including the total supply power (P_{sup}), the amount of actual energy exchange between servo drives (P_{exc}) and the brake resistor dissipation power (P_{chp}).

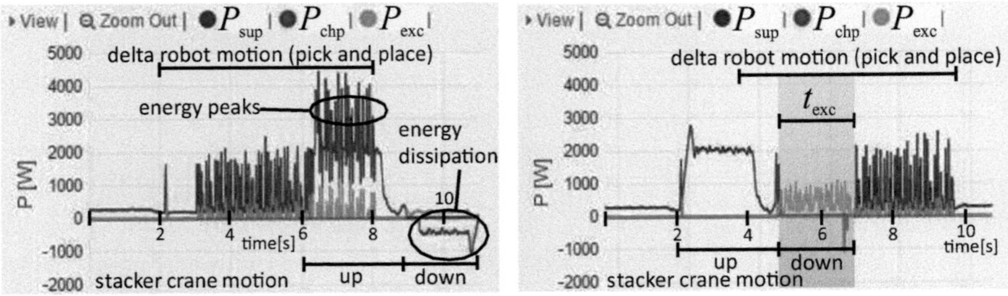

(a) Initial motion sequence (inefficient) (b) Optimized motion sequence (efficient)

Fig. 4: Web-dashboard for cloud-based energy monitoring

The data **Analysis** results in Fig. 4(a) illustrate that the initial process sequence is not energy-efficient. Energy recuperated by the stacker crane during downwards motion is not effectively used by other system modules and must be dissipated in each cycle. Based only on visualization of the results, the process can be manually optimized e. g. by adjusting a cyclic start time delay on the delta robot. Automatic adjustments will be considered for future works. As a result Fig. 4(b) shows, that the recuperated energy of the stacker crane is consumed during the delta robot pick and place motion (t_{exc}), increasing the energy exchange within the DC link, minimizing the energy supply by around 11% (externally measured), distinctly reducing the energy peak loads and avoiding energy dissipation almost entirely. The presented Data Analytics Infrastructure enables a deeper process understanding and hence forms a sound information basis for the selection of different process and plant optimization strategies.

5 Conclusion and Future Work

To cope with the challenges of factory plant and process digitalization including handling of Big Data, a Data Analysis Infrastructure has been implemented in a private cloud environment to enable processing of large data streams. Build on open-source software, the architecture unifies heterogeneous data streams and provides a platform for data-driven industrial analytics applications. The functionality and efficiency were demonstrated using the exemplary use-case of energy monitoring and optimization for a laboratory model factory. This infrastructure enables the acquisition of large volume data streams from the field layer (e. g. from PLCs, fieldbuses and sensors) and the computation of processing-intensive models in the wide area of mechatronic systems. Depending on different use-cases, including condition monitoring and predictive maintenance, future work will focus on the implementation of domain-specific analytical models and data-driven methods from the field of Data Science.

Acknowledgment

The authors like to thank Lenze SE for providing the control cabinet (including power electronics components and PLCs) and thus making this research possible.

References

1. Weyer, S., Schmitt, M., Ohmer, M., Gorecky, D.: Towards Industry 4.0-Standardization as the crucial challenge for highly modular, multi-vendor production systems. IFAC-PapersOnLine, vol. 48, num. 3, pp. 579–584 (2015).
2. Qin, S.J.: Process data analytics in the era of big data. AIChE Journal, vol. 60, num. 9, pp. 3092–3100 (2014).
3. O'Donovan, P., Leahy, K., Bruton, K., O'Sullivan, D.T.J.: An industrial big data pipeline for data-driven analytics maintenance applications in large-scale smart manufacturing facilities. Journal of Big Data, vol. 2, num. 1, pp. 1–22 (2015).
4. Lee, J., Lapira, E., Bagheri, B., Kao, H.: Recent advances and trends in predictive manufacturing systems in big data environment. Manufacturing Letters, vol. 1, num. 1, pp. 38–41 (2013).
5. Felser, M.: Real-time ethernet industry prospective. In Proceedings of the IEEE, vol. 93, num. 6, pp. 1118–1129 (2005).
6. Steiner, W., Poledna, S.: Fog computing as enabler for the Industrial Internet of Things. e & i Elektrotechnik und Informationstechnik, vol. 133, num. 7, pp. 310–314 (2016).
7. Windmann, S., Maier, A., Niggemann, O., Frey, C., Bernardi, A., Gu, Y., Pfrommer, H., Steckel, T., Krüger, M., Kraus, R.: Big Data Analysis of Manufacturing Processes. Journal of Physics: Conference Series of 12th European Workshop on Advanced Control and Diagnosis, vol. 659, num. 1, pp. 1–12 (2015).
8. Schleipen, M., Kühnert, C., Okon, M., Henßen, R., Bischoff, T.: Mobile Monitoring und smarte Datenanalyse basierend auf offenen Standards. VDI/VDE-Gesellschaft Meß- und Automatisierungstechnik -GMA-, Düsseldorf: Automation (2015).
9. Russom, P.: Big data analytics. TDWI best practices report (2011).
10. Yue, X., Cai, H., Yan, H., Zou, C., Zhou, K.: Cloud-assisted industrial cyber-physical systems: an insight. Microprocessors and Microsystems, vol. 39, num. 8, pp. 1262–1270 (2015).
11. Hashem, I.A.T., Yaqoob, I., Anuar, N.B., Mokhtar, S., Gani, A., Khan, S.U.: The rise of "big data" on cloud computing: Review and open research issues. Information Systems, vol. 47, pp. 98–115 (2015).
12. Hallmans, D., Sandström, K., Nolte, T., Larsson, S.: Challenges and opportunities when introducing cloud computing into embedded systems. In Proceedings of the 13th IEEE International Conference on Industrial Informatics (INDIN), pp. 454–459 (2015).
13. Dijcks, J.-P.: Oracle: Big Data for the Enterprise. An Oracle White Paper, Redwood Shores, CA: Oracle Corporation (2013).
14. White, M.: Digital workplaces: vision and reality. Business information review, Sage Publications, vol. 29, num. 4, pp. 205–214 (2012).
15. Vick, A., Horn, C., Rudorfer, M., Krüger, J.: Control of robots and machine tools with an extended factory cloud. In Proceedings of the IEEE World Conference on Factory Communication Systems (WFCS), pp. 1–4 (2015).

16. Kretschmer, F., Borisov, A., Pöschko, R.: Chemnitz, M., Vick, A.: Steuerungstechnik aus der Cloud - Anwendungsszenarien für cloudbasierte Produktion im Rahmen des Forschungsprojekts pICASSO. wt Werkstattstechnik online, vol. 106, num. 5, pp. 308–313 (2016).
17. Jin, X., Wah, B.W., Cheng, X., Wang, Y.: Significance and challenges of big data research. Big Data Research, vol. 2, num. 2, pp. 59–64 (2015).
18. Lee, J., Kao, H., Yang, S.: Service innovation and smart analytics for industry 4.0 and big data environment. Procedia CIRP, vol. 16, pp. 3–8 (2014).
19. The Apache Hadoop Project. http://hadoop.apache.org/core/, Last Access: 03/02/2017
20. White, T.: Hadoop: The Definitive Guide. O'Reilly Media Inc., Sebastapol, CA (2009).
21. Shvachko, K., Hairong, K., Radia, S., Chansler, R.: The Hadoop Distributed File System. In Proceedings of the 26th IEEE Symposium on Mass Storage Systems and Technologies (MSST), pp. 1–10 (2010).
22. Stonebraker, M.: SQL databases v. NoSQL databases. Communications of the ACM, vol. 53, num. 4, pp. 10–11 (2010).
23. Apache Kafka. https://kafka.apache.org/, Last Access: 03/02/2017
24. Apache Storm. http://storm.apache.org/, Last Access: 03/02/2017
25. Apache HBase. https://hbase.apache.org/, Last Access: 03/02/2017
26. Apache Solr. http://lucene.apache.org/solr/, Last Access: 03/02/2017
27. Öltjen, J., Beckmann, D., Hansen, C., Maurer, I., Kotlarski, J., Ortmaier, T.: Integrated Parameter Management Concept for Simplified Implementation of Control, Motion Planning, and Process Optimization Methods. Applied Mechanics & Materials, vol. 840, pp. 114–122, Trans Tech Publications, Schweiz, DOI: 10.4028/www.scientific.net/AMM.840.114 (2016).
28. Hansen, C., Öltjen, J., Meike, D., Ortmaier, T.: Enhanced Approach for Energy-Efficient Trajectory Generation of Industrial Robots. Proceedings of the 8th IEEE International Conference on Automation Science and Engineering, pp. 1–7 (2012).

Normal Operation Input Signals for Parameter Estimation in Underactuated Structures

Gunnar Borchert[1], Jonas Diekmeyer[2], Konrad Bild[2], and Annika Raatz[1]

[1] Leibniz Universität Hannover,
Institute of Assembly Technology,
An der Universität 2, 30823 Garbsen
{borchert, raatz}@match.uni-hannover.de,
[2] Leibniz Universität Hannover, student

Abstract. This paper addresses challenges of parameter estimation of an arbitrary object which is manipulated by an underactuated handling system. In the present scenario, a robot is extended with a passive orientation device. Since the passive joints are steered by energy control, knowledge of the inertial parameters of the gripped object must be obtained. For this purpose, an evaluation process is shown to find excitation inputs that are based on normal operation motion profiles. The general applicability of the excitation is then demonstrated along with an optimization to improve the excitation of the passive joints which yields a better estimation. Since it is difficult to obtain acceleration signals, the influence of their accuracy on the estimates is additionally illustrated. The article closes with the identification of future developments.

Keywords: Handling, Underactuated Structure, Parameter Estimation

1 Introduction

In the recent past, it was shown that mechanical structures with passive revolute joints can be used to enhance the mobility of existing handling robots [1]. In Fig. 1 the kinematics of a 4-dof industrial delta robot is depicted that is equipped with a passive 2-dof orientation device to obtain a full orientation workspace. Hence, $\mathbf{q}_\mathrm{a} = (q_1, \dots, q_4)^\mathrm{T}$ denotes the four active prismatic and revolute joints of the delta robot and $\mathbf{q}_\mathrm{p} = (q_5, q_6)^\mathrm{T}$ comprises the attached two passive revolute joints of the orientation device. Additional dof usually imply the installation of additional motors as commonly done in industry [2] or research [3]. In this context, the effort of excluding additional motors is driven by a reduction in costs, mechanical stress and power consumption or the enhancement of the reliability [4]. This will in turn give rise to some challenges, like difficult control and design properties. A possible control strategy is presented in [1], where the proposed robot system is steered by energy-based swing-up control [5] and dynamic coupling [6]. However, for this purpose, calculating the energy of the passive structure with high accuracy becomes essential so that it will be necessary to obtain knowledge of the respective model parameters that influence the motion of the passive joints.

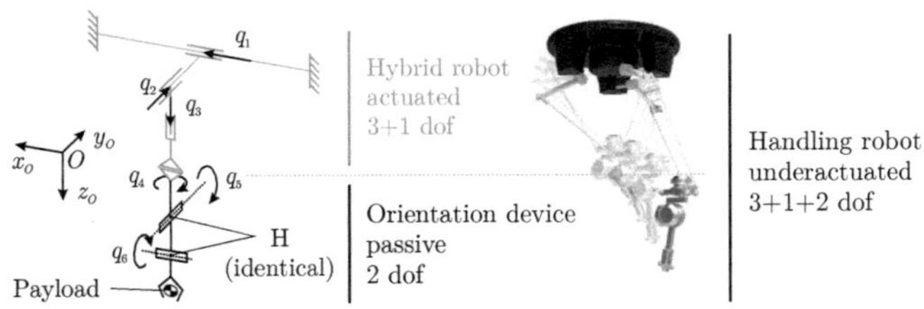

Fig. 1. Kinematics of a delta robot which is extended with two passive joints.

However, the parameters of the robot model will be changed whensoever an object is gripped. For this purpose, our work contributes to the automated estimation of such unknown payload inertia. By means of a generative framework, the linear least-square solution is combined with a particle swarm optimization (PSO) algorithm to find an optimal trajectory setting for the excitation of the passive joints. To this, the continuity of different normal operation motions as well as the quality of the acceleration profile and their influence on the estimation is presented.

At first, the superior evaluation framework is introduced in Sec. 2. This is followed by a brief introduction on linear regression and parameter scaling in Sec. 3 in order to analyze different input signals. In addition, the applicability of the excitation is subsequently shown. In Sec. 3.4, the rudimental idea of an optimization strategy is proposed to improve the estimation process. Finally, possible future investigations are summarized in Sec. 4.

2 Process design

The evaluation framework is depicted in Fig. 2. Input parameters are the excitation $\mathbf{q}_\mathrm{a}$, which is defined by the move distance Δs and path $s(t)$ of the wrist center H as well as its duration T and move direction defined by θ_S. Other inputs are the accurate values $\mathbf{p}_0$ of the payload inertia. The output is a set of estimates $\hat{\mathbf{p}}$ of the payload inertia $\mathbf{p}_\mathrm{p}$ along with the squared estimation error $\mathbf{e}^2$. In general, a parametric model that includes Coulomb and viscous friction is used in order to emulate a real system behavior and to generate pseudo data. This is subsequently discretized to further imitate an encoder signal with 16-bit resolution. Assuming that the frictional behavior deviates from the real system, an error signal e_Fric can be mapped onto it. The emulation is also used for a sub-process to optimize the motion profile with respect to the rate of excitation by means of an object function f, cf. Sec. 3.4. The motion is included in a typical pick-and-place handling process, as shown in Fig. 2. To minimize f, the PSO is applied. The PSO is not part of this article, comprehensive information about it and its implementation is given by *Marini* and *Walzek* in [7]. The PSO varies a parameter set $\mathbf{p}_\mathrm{s} = (\theta_\mathrm{s}, \Delta s)^\mathrm{T}$ which includes θ_s and Δs in the first instance. The angle θ_s defines the direction of $s(t)$ in the x_0y_0-plane using the rotation matrix $\mathbf{R}_\mathrm{z}$ about the z_0-axis, $\mathbf{q}_\mathrm{a} = \mathbf{R}_\mathrm{z}(\theta_\mathrm{s})[s(t), \mathbf{0}, \mathbf{0}]^\mathrm{T}$. Of course, $\mathbf{p}_\mathrm{s}$ can be adapted for other parameter sets and different motion profiles. The found solution is then used to obtain the final parameter set $\hat{\mathbf{p}}$. In summary, the framework is used to imitate the real system for parameter estimation of the payload inertia. This is done by means of a trajectory $s(t)$. Due to the influence of the parametrization of $s(t)$ on the quality of the estimates, the PSO is applied to improve the outcome of the estimation.

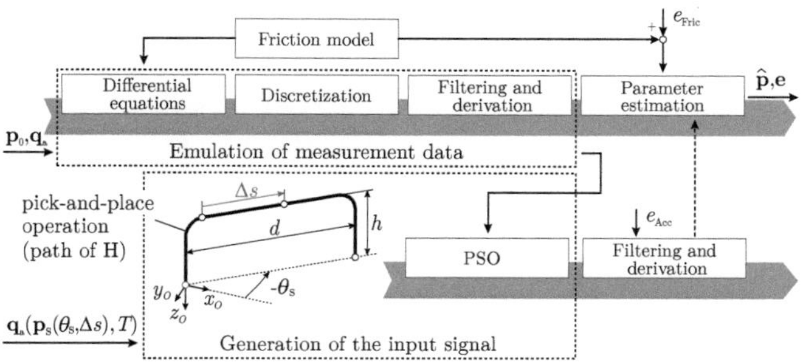

Fig. 2. Scheme of the evaluation framework.

3 Parameter estimation

Parameter estimation is a topic in system identification by which a previously modeled system behavior is attuned to its real counterpart. In this context, the use of non-parametric (e.g. transfer functions) and parametric models (e.g. differential equations) are distinguished. In the latter case, the parameters are explicitly included and an initial setting can be assumed from theoretical modelling. By contrast, non-parametric models do not give an insight into the system design, but rather provide an abstract description of its dynamic behavior [9].

Since our approach includes a parametric model the parameters $\mathbf{p}_{\mathrm{p}}$ must be estimated explicitly. By considering a spatial system, they comprise the three linear independent parameters $m_{\mathrm{p}}l_{\mathrm{p,x}}$, $m_{\mathrm{p}}l_{\mathrm{p,y}}$ and $m_{\mathrm{p}}l_{\mathrm{p,z}}$, which are composed of the mass and the distances from point H to the object's center of gravity. Additional components are $J_{\mathrm{p,xx}}$, $J_{\mathrm{p,xy}}$, $J_{\mathrm{p,xz}}$, $J_{\mathrm{p,yy}}$, $J_{\mathrm{p,yz}}$ and $J_{\mathrm{p,zz}}$ of the inertia matrix. The measured variables for the estimation are $\mathbf{q}_{\mathrm{p}}$ and their derivatives.

3.1 Regression and parameter scaling

The implemented method for parameter estimation is based on the well-known linear least-square solution, which aims on minimizing the square sum $\mathbf{e}^2$ of the residuals $\mathbf{e} = \mathbf{U} \cdot \mathbf{p} - \mathbf{y}$ that is $\min_{\mathbf{p}} \|\mathbf{U} \cdot \mathbf{p} - \mathbf{y}\|_2^2$. Here $\mathbf{U}$ denotes the regression matrix, $\mathbf{p}$ the vector of unknown linear independent inertial parameters (of the payload) and $\mathbf{y}$ includes the measured output data and defines the number of $\dim(\mathbf{y})$ measurements. The estimation is done by calculating the pseudo-inverse $\mathbf{U}^{+} = (\mathbf{U}^{\mathrm{T}}\mathbf{U})^{-1}\mathbf{U}^{\mathrm{T}}$ to solve $\mathbf{p} = \mathbf{U}^{+}\mathbf{y}$. However, the parameters could have vastly different magnitudes which inhibits a comparison of the singular values of $\mathbf{U}$, as it will be done in Sec. 3.4. Consequently, parameter scaling [10] is used which is a right multiplication of $\mathbf{U}$ by a weighting matrix $\mathbf{W}$ [11] so that $\tilde{\mathbf{U}} = \mathbf{UW}$ and $\tilde{\mathbf{p}} = \mathbf{W}^{-1}\mathbf{p}$. To this, $\mathbf{W}$ can be determined by column scaling, which is '*...the most common approach towards parameter weighting...*' and '*...does not require a priori statistical information.*' [10]. The weighting matrix then becomes $\mathbf{W} = \mathrm{diag}(d_1, \ldots, d_n)$, with $d_i = \|c_i\|^{-1}$ if $\|c_i\| \neq 0$ and $d_i = 1$ if $\|c_i\| = 0$, where $i = 1 \ldots \dim(\mathbf{p})$ denotes the column c_i of $\mathbf{U}$.

3.2 Input signals

The choice of the excitation signal is crucial to get sufficient identification data for the parameter estimation. To this end, step or pseudo-random binary sequence (PRBS, white noise) signals are known, where the latter is established to excite the dynamic behavior of linear processes [9] so that this behavior can be modeled by means of the measured output data. In this context, *Isermann* [12] stated some general characteristics which indicate the suitability of an input signal. Its generation must ideally be simple, repeatable and mathematically describable for the identification method. Moreover, it must excite the dynamic behavior of the system sufficiently well and finally be applicable to the process. In other words, its actuation must be feasible. Especially the latter statement suggests that ideal step or white noise signals are difficult to realize if a real robot system is used that is determined by external disturbances, control inaccurancies or limited dynamics. In this case, normal operation inputs may be a solution to the identification problem. In the following, normal operation inputs are defined to be motion profiles that are commonly used for the respective handling robot. Such profiles are also advantageous because they are simply repeatable and they instantly bound the motion within valid states. This avoids system failure and will protect against mechanical damages [13]. A comprehensive overview of different motion profiles is given by *Biagotti* et. al in [14], whereas Tab. 3.2 comprises an abstract by taking different profile characteristics into account. Their effects on the estimation process are subsequently analyzed with the first sub-process. To this end, e_{Fric} was realized by means of a second friction model that deviates from the original model by $< 2\,\%$ (assumption). For the sake of clarity, a reduced 2-dof model with an actuated joint q_1 and a passive joint q_5 is used in Sec. 3.2.

In the first instance, the description is independent from the parameter setting. The continuity is consulted since first tests indicated its dependence on the estimation error, as depicted in Fig. 3. Here, $e^2_{1,\max}$ scales with the jerk ($e^2_{1,\max} = 157\,\mathrm{rad}^2\,\mathrm{s}^{-4}$ at $q^{[3]}_{1,\max} = 800\,\mathrm{m\,s}^{-3}$ and $e^2_{1,\max} = 354\,\mathrm{rad}^2\,\mathrm{s}^{-4}$ at $q^{[3]}_{1,\max} = 1200\,\mathrm{m\,s}^{-3}$) but diminishes significantly if the profile is continuous ($e^2_{1,\max} = 35\,\mathrm{rad}^2\,\mathrm{s}^{-4}$ at $q^{[3]}_{1,\max} = 1440\,\mathrm{m\,s}^{-3}$). Likewise, it could be observed for the exponential and 15-segement profiles. The estimation error can be traced back to deviations if the position signal is discretized and filtered. In particular, this has an inherently high impact on the second derivative which is used for the estimation, cf. Sec. 3.5. In view of advantageous error properties, trajectory

Table 1. Normal operation motion profiles, which are commonly used for handling systems. The position (q), velocity ($\dot{q}_1$), acceleration ($\ddot{q}_1$) and jerk ($q_1^{[3]}$) signal is continuous (✓) or discontinuous (x). In addition, the (dis-)continuity of the respective signals at their initial (index i) and end (index e) points is marked.

No.	Type	q_1	$\dot{q}_1$	$\ddot{q}_1$	$q_1^{[3]}$	$q_{1,\mathrm{ie}}$	$\dot{q}_{1,\mathrm{ie}}$	$\ddot{q}_{1,\mathrm{ie}}$	$q_{1,\mathrm{ie}}^{[3]}$
1	Polynomial (7th order)	✓	✓	✓	✓	✓	✓	✓	✓
2	Polynomial (Split polynoms, 4-3-4)	✓	✓	✓	x	✓	✓	✓	x
3	Harmonic	✓	✓	✓	✓	✓	✓	x	✓
4	Exponential	✓	✓	✓	✓	✓	✓	✓	✓
5	Fourier series	✓	✓	✓	x	✓	✓	✓	x
6	7-segments trajectory	✓	✓	✓	x	✓	✓	✓	x
7	15-segments trajectory	✓	✓	✓	✓	✓	✓	✓	✓

No. 1 was chosen for further investigations in the following sections. In addition, the implementation is simple since the trajectory comprises a single polynom for each derivative along with its initial and final state as well as the duration. In comparison, segmented trajectories are defined by cases which is typically more extensive to implement.

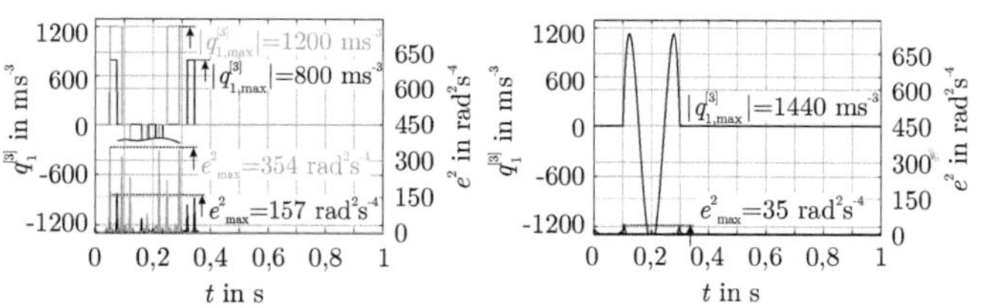

Fig. 3. Residuals which result from the motion profiles No. 6 (left) and No. 1 (right).

3.3 Applicability

With regard to Fig. 1, the input signal $\mathbf{q}_\mathrm{a}$, which moves point H along $s(t)$, must be chosen as to cause a change in the measures of $\mathbf{q}_\mathrm{p}$. This can exemplarily be assessed by interpreting the coupling index $\rho = \mathrm{sqrt}(\det(\mathbf{M}_\mathrm{pp}^{-1}\mathbf{M}_\mathrm{pa}\mathbf{M}_\mathrm{pa}^\mathrm{T}\mathbf{M}_\mathrm{pp}^{-\mathrm{T}}))$ [15]. The coupling index depends on the kinematic structure and allows for rating the dynamic interaction between active and passive joints due to inertial effects. As given, the index results from the collocated partial feedback linearization of the equations of motion of the system which are known as the 2nd order nonholonomic constraints $\ddot{\mathbf{q}}_\mathrm{p} = -\mathbf{M}_\mathrm{pp}^{-1} \cdot \mathbf{h}_\mathrm{p} - \mathbf{M}_\mathrm{pp}^{-1} \cdot \mathbf{M}_\mathrm{pa} \cdot \ddot{\mathbf{q}}_\mathrm{a}$ [16]. $\ddot{\mathbf{q}}_\mathrm{p}$ and $\ddot{\mathbf{q}}_\mathrm{a}$ denote the acceleration vectors of the passive and active joints, respectively. $\mathbf{M}_\mathrm{pp}$ and $\mathbf{M}_\mathrm{pa}$ are components of the mass matrix and $\mathbf{h}_\mathrm{p}$ comprises damping, stiffness and Coriolis effects of the passively coupled links. ρ is thus a matter of singular values of $\mathbf{M}_\mathrm{pp}^{-1}\mathbf{M}_\mathrm{pa}$ which signify its invertability or rather the influence of $\ddot{\mathbf{q}}_\mathrm{a}$ on $\ddot{\mathbf{q}}_\mathrm{p}$. The coupling index is presented in Fig. 4 (a) for the given orientation device acting in $x_0y_0z_0$-direction. It shows that each passive joint can be excited, because $\rho \neq 0$. Only if ρ equals zero, $\ddot{\mathbf{q}}_\mathrm{a}$ will lose the influence on $\ddot{\mathbf{q}}_\mathrm{p}$. Since the motion range of q_3 is typically small for handling robots, diagram (b) shows the singularities, $\rho = 0$, when q_3 remains constant. The excitation of the passive joints should then be restricted so that q_5 and q_6 are apart from mechanical restrictions and singularities. The latter becomes crucial, since the identification process must remain stable to the effect that input and output signals are unambiguous assigned, cf. [9]. In conclusion, it can be stated that it is possible to excite all passive joints by means of the available inputs. However, this approach shows the general possibility of exciting the passive joints and a quantitative comparison between them. The analysis cannot be used to find a lower limit value that yields a sufficient excitation.

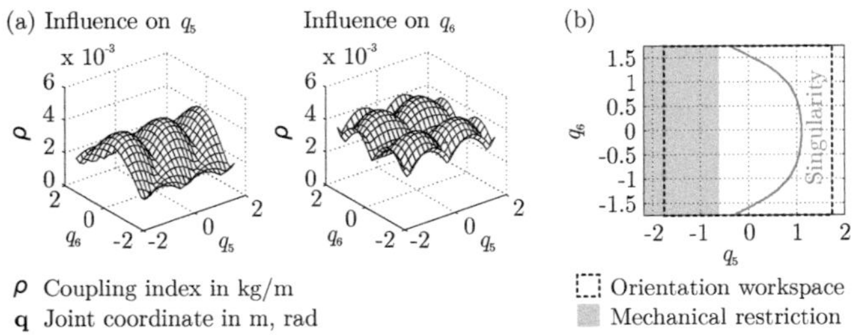

Fig. 4. Coupling index (a) and singularities (b) of the orientation device.

3.4 Optimization

As already mentioned before, the passive joints must be excited, since they provide the basis for the output $\mathbf{y}$. Due to the fact that the motion profile is just a straight line, it can happen that one or more joints are not sufficiently excited. This phenomenon occurs for the inital configuration where the parameters are assigned to $\mathbf{p}_{\mathrm{s},0} = [0, 0.5]^\mathrm{T}$. The duration was set to 0.5 s. In this case, joint q_6 cannot be controlled which can be deduced from $\dot{\mathbf{q}}_6 \approx 0$, cf. Fig. 5.

With this in mind, the optimization aims at improving the excitation of $\mathbf{q}_\mathrm{p}$. For this purpose, an evaluation criterion is needed to score the quality of the estimation. Suitable criteria are often based on observability indexes where each index comprises a different set of the i singular values σ_i of $\mathbf{U}$, $\sigma_i \neq 0$. In [17] and [10], the physical meaning of such indexes is given and an interrelation is discussed. For inertial parameters identification, *Gautier* presented the object function $f(\mathbf{q}, \dot{\mathbf{q}}) = \lambda_1 \kappa(\mathbf{U}(\mathbf{q}, \dot{\mathbf{q}})) + \lambda_2 \mathrm{S}(\mathbf{q}, \dot{\mathbf{q}})$ and proved its benefits for optimal motion planning [18],[19]. Here, the condition $\kappa(\mathbf{U}) = \max(\sigma_i)/\min(\sigma_i)$ represents the extreme singular values of the regressor matrix. λ_1 and λ_2 are weighting factors and $\mathrm{S} = \max(|U|)/\min(|U|)$, with $\min |U| \neq 0$, describes the equilibration of the entries of $\mathbf{U}$. However, it is stated in [10] that '*... small singular values σ_j greatly perturb the estimates, because the weigthing is* $1/\sigma_j$.' For this reason, we use a modified object function in Eq. 1 to maximize $\min(\sigma_i)$ in addition. The first and second summand becomes zero when the entries of $\kappa(\mathbf{U})$ and $\mathbf{U}$, respectively, are well equilibrated, whereas the third summand becomes zero for large σ_i. For the moment, the weighting factors are neglected and equal one. However, λ_2 should be reduced to damp the influence of small entries in $\mathbf{U}$ on the evolution of f. To this, Schröer [20] is cited in [10], who suggests the heuristic $\kappa(\mathbf{U}) < 100$ so that $\mathbf{U}$ is said to be well-conditioned.

$$f(\mathbf{q}, \dot{\mathbf{q}}, \ddot{\mathbf{q}}) = \lambda_1 \left(\frac{\max(\sigma_i)}{\min(\sigma_i)} - 1 \right) + \lambda_2 \left(\frac{\max(|U|)}{\min(|U|)} - 1 \right) + \lambda_3 \left(\frac{1}{\min(\sigma_i)} \right) \quad (1)$$

With respect to f, the parameter evolution is shown in Fig. 5 (left) for $\theta_\mathrm{s} \in [0, 360[$ and $\Delta s \in [0.1, 0.5]$. The parametrization was assumed to be meaningful for the desired handling operation. In addition, boundary conditions of $\mathbf{q}_\mathrm{p}$ are taken into account. For this purpose, costs are assigned to infinity to neglect the respective parameter set which yields angular positions beyond singularities or mechanical restrictions, as stated in Sec. 3.3. One suitable minimum of the object function can be found for the parameter set $\mathbf{p}_{\mathrm{s},0} = (60.2, 0.16)^\mathrm{T}$. f decreases from 2.04×10^8 to 3188. At the same time $\min(\sigma_i)$ can be improved from 0.012 before to 0.051 after optimization. Likewise, $\kappa(\mathbf{U})$ can be lowered from 144.06 before to 33.25 after optimization so that it is below 100. The result shows lastly a much better excitation of both joints $\mathbf{q}_\mathrm{p}$ together, see Fig. 5

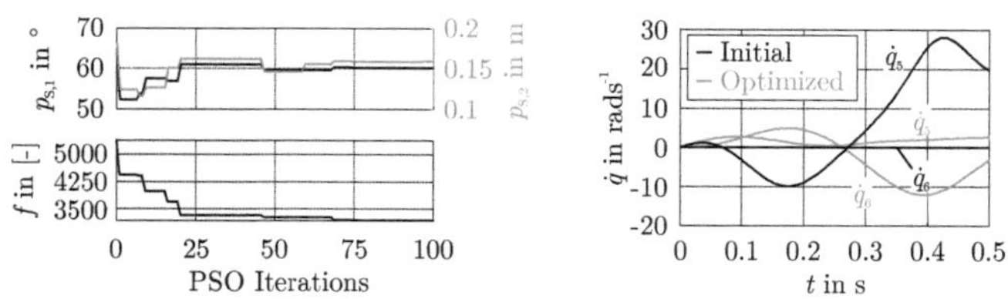

Fig. 5. Evolution of the parameters and the object function (left); Excitation of the passive joints before and after PSO (right).

3.5 Influence of inaccurate acceleration data

To investigate the influence of the quality of the acceleration derivative on the estimation, we mapped an error function $e_{\mathrm{Acc}}(\ddot{\mathbf{q}}_\mathrm{p})$, which is expressed as a percentage of the acceleration profile $\ddot{\mathbf{q}}_\mathrm{p}$, onto $\ddot{\mathbf{q}}_\mathrm{p}$ to analyze their influence, cf. Fig. 2. This is due to the fact that the acceleration profile is neither easy to measure nor easy to calculate [8], but it strongly influences the accuracy of the estimation results. In the following, the estimation parameters $\mathbf{p}_\mathrm{p}$ are assigned to $m_\mathrm{p} l_{\mathrm{p,x}} = 0.0132\,\mathrm{kg\,m}$, $m_\mathrm{p} l_{\mathrm{p,y}} = -0.0009\,\mathrm{kg\,m}$ and $m_\mathrm{p} l_{\mathrm{p,z}} = -0.0018\,\mathrm{kg\,m}$, $J_{\mathrm{p,xx}} = 0.0004\,\mathrm{kg\,m^2}$, $J_{\mathrm{p,yy}} = 0.0002\,\mathrm{kg\,m^2}$ and $J_{\mathrm{p,zz}} = 0.0002\,\mathrm{kg\,m^2}$. The moments of deviation are $J_{\mathrm{p,yz}} = -9 \times 10^{-7}\,\mathrm{kg\,m^2}$, $J_{\mathrm{p,xy}} = -7 \times 10^{-7}\,\mathrm{kg\,m^2}$ and $J_{\mathrm{p,xz}} = 2.7 \times 10^{-6}\,\mathrm{kg\,m^2}$. The parameters result from an arbitrarily chosen test object. The least-square solution is illustrated in Fig. 6, where e_{ACC} is increased from 0 % to 10 %. The results are based on the optimized motion profile and depict the relative deviation of the estimates $\Delta\mathbf{p}_{\mathrm{p}}$ from the accurate values so that $\Delta\mathbf{p}_{\mathrm{p}} = \frac{\mathbf{p}_{\mathrm{p}}-\mathbf{p}_0}{\mathbf{p}_0} \cdot 100\,\%$.

In the present simulation, a uniformly continuous reduction of the values from ≈0 % to ≈ − 150 % can be deduced from Fig. 6 (left). By contrast, the parameters in Fig. 6 (right) alternates between ≈280 % to ≈ − 350 % for $J_{\mathrm{p,xy}}$ and between ≈130 % to ≈ − 50 % for the remaining moments of deviation. The high variation and the alternating sign result from small values of the moments of deviation which are in fact typically much smaller than the principal moments of inertia. In general, it can be stated from different simulations, that the higher the values, the better the estimation performs.

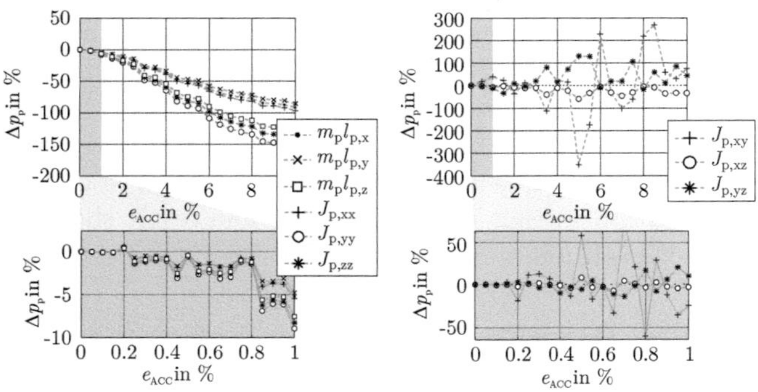

Fig. 6. Influence of the accuracy of $\ddot{\mathbf{q}}_{\mathrm{p}}$ on the estimation

4 Conclusions and future work

The article addresses the parameter estimation of an object that is manipulated by a passive orientation device. The device is attached to an industrial robot and steered by energy control and inertial coupling. Thus, the inertials of the gripped object were obtained by considering the motion of the passive axes. The joints were excited by means of normal operation motions where the feasibility of the excitation was investigated. It was shown that the estimation error scales with the jerk and is related with its discontinuity. To improve the excitation and thus the estimation, an optimization strategy was introduced which allows for the adaptation of the profile parameters by minimizing an object function. In addition, the influence of inaccurate acceleration measurements was illustrated which diminishes the estimation results similar to the size of the values. Finally, practical tests need to be carried out. This includes different objects as well as inaccurate gripping. In addition, the inaccuracies due to the acceleration derivative should be reduced by means of improved filter techniques.

References

1. Borchert, G., Raatz, A., 2016, A new method for combining handling systems with passive orientation devices, In: CIRP Annals - Manufacturing Technology, Elsevier B.V., 2016, Vol. 65/1, pp. 49-52.
2. Fanuc Robotics America Corporation. Datasheet: FANUC Robot M-3iA. FRA-24/02/2014, Online available via: www.fanucrobotics.com [cited on 14.07.15].
3. Liu, N., Wu, J., 2014, Kinematics and Application of a Hybrid Industrial Robot - Delta-RST, In: Sensors&Transducers, Vol. 169/4, April 2014, pp. 186-192.
4. Borchert, G., Raatz, A., 2015, An Analysis Process to Improve the Mobility of a Parallel Robot for Assembly Tasks, 14th World Congress in Mechanism and Machine Science, 25-30 Oct., 2015, Taipei, Taiwan.
5. Astroem, K.-J., Furuta, K., 2000, Swinging up a pendulum by energy control, Journal Automatica, 36/2:287-295.
6. De Luca, A., Iannitti, S., Mattone, R., Oriolo, G., 2002, Underactuated manipulators: Control properties and techniques, Machine Intelligence and Robotic Control, Vol. 4/3, pp. 113-125.
7. Marini, F., Walczak, B., 2015, Particle swarm optimization (PSO). A tutorial, In: Chemometrics and Intelligent Laboratory Systems, Vol. 149, Part B, 15 December 2015, pp. 153-165.
8. Gautier, M., Khalil, W., 1991, Exciting Trajectories for the Identification of Base Inertial Parameters of Robots, In: Proc. of the 30th Conference on Decision and Control, Brighton, England, Dec., 1991, pp. 494-499.
9. Isermann, R., 2008, Mechatronische Systeme. Grundlagen, Springer Berlin Heidelberg, 2008, DOI 10.1007/978-3-540-32512-3
10. Siciliano, B., Khatib, O., 2008, Springer Handbook of Robotics, Springer Berlin Heidelberg, 2008, DOI 10.1007/978-3-540-30301-5
11. Lawson, C.L., Hanson, R.J., 1995, Solving Least Squares Problems, Philadelphia : SIAM, c1995, ISBN-13: 978-0898713565
12. Isermann, R., 1988, Identifikation dynamischer Systeme 1: Grundlegende Methoden, Springer Berlin Heidelberg, 1992, DOI 10.1007/978-3-642-84679-3
13. Nelles, O., 2001, Nonlinear System Identification, Springer Berlin Heidelberg, 2001, DOI 10.1007/978-3-662-04323-3
14. Biagotti, L., Melchiorri, C. (2008): Trajectory Planning for Automatic Ma-chines and Robots, ISBN: 978-3-540-85628-3.
15. Bergerman, M., Lee, C., Xu, Y., 1995, Dynamic Coupling of underactuated Manipulators, In: Proc. of the 4th IEEE Conference on Control Applications, Albany, USA, Sep. 1995, pp. 500-505.
16. Allgöver, F., Gilles, E.D., 1993, Nichtlinearer Reglerentwurf auf der Grundlage exakter Linearisierungstechniken, VDI Berichte Nr. 1026, 1993, pp. 209-234.
17. Sun, Y., Hollerbach, M., 2008, Observability Index Selection for Robot Calibration, In: IEEE Int. Conference on Robotics and Automation, Pasadena, USA, May 19-23, 2008, pp. 831-836.
18. Gautier, M., 1992, Optimal Motion Planning For Robots Inertial Parameters Identification, In: Proc. of Decision and Control, 16-18 Dec., Tucson, 1992, pp. 70-73.
19. Gautier, M., Khalil, W., 1992, Exciting Trajectories for the Identification of Base Inertial Parameters of Robots, In: The Int. Journal of Robotics Research, Vol. 11/4, Aug., 1992, pp. 362-375.
20. Schroer, K., 1993, Theory of kinematic modelling and numerical procedures for robot calibration, In: Robot Calibration, ed. by R. Bernhardt, S.L. Albright (Chapman Hall, London 1993), pp. 157-196.

Joining Linked Micro Formed Parts through Tolerance Field Widening and Synchronization

Ann-Kathrin Onken[1a], Patrick Rückert[1b], Christian Perl[1], Kirsten Tracht[1]

[1]University of Bremen, Bremen Institute for Mechanical Engineering (bime), Bremen, Germany

[a]onken@bime.de, [b]rueckert@bime.de

Abstract. Metal forming and joining of micro components is related to a high number of requirements. Especially small manufacturing tolerances, scaling effects and mechanical handling efforts are still challenging. Due to these factors, production processes from macro components cannot be transferred easily to micro range. One way to address these issues is the manufacturing of linked micro parts, which results in the possibility of enabling facilitated conditions for the handling and positioning during micro-forming and joining processes. However, the usage of linked parts leads to new challenges, especially regarding the synchronization of process chains for joining. Therefore, concepts are required, that consider the challenges of process times, matching of linked components and discharging of scrap parts. All in all, this paper gives an insight of current challenges in synchronizing process chains for joining micro formed linked parts and presents solution statements to address them.

Keywords: Joining, Micro Machining, Tolerance Field Widening

1 The Relevance of Micro Production and its Challenges

Micro parts are used in miscellaneous industries like medical and electrical engineering. Micro production is defined as the production of parts that are in sub-millimeter range in two dimensions [1]. Due to the occurrence of size effects, the production and handling of micro parts are still challenging.

Size effects arise while scaling geometric dimensions of piece parts. These effects could be categorized in accordance to density, shape and microstructure [2,3]. The handling of separated micro parts is for example influenced by the ratio of volume to surface and occurring adhesion effects as well as the sensitivity of micro parts [4]. The production of linked parts enables an up scaling of the produced micro parts. In this case a belt or the original material like foil or wire is used for interconnecting parts. The links are not released unless it is mandatory [5]. This facilitates a handling of several micro parts as one macro part and avoids the occurrence of disturbing size effects. Due to the connection of the micro parts a higher knowledge of the process is required for a synchronization of the paced production processes within the joining point [6]. In accordance to [5], linked parts could be classified into three basic types. A ladder type, a line type and comb linked parts as a hybrid, combined type of the first mentioned.

Within the subproject C5 "Linked Parts" of the Collaborative Research Center 747 (CRC 747) "Micro Cold Forming - Processes, Characterization, Optimization" at the University of Bremen the process synchronization and control of linked parts is investigated. The configuration of the process is investigated by the example of cups and spheres that are produced, linked and joined in a synchronization scenario. An identification of trends within the forming process could be used for an adjustment of the nominal value through the linked forming of parts. This enables a higher outcome of well joined parts as well as a longer use of tools because the appearing wear can be utilized. All in all this facilitates a change of the paradigms in micro production. As depicted in figure 1, the new paradigm is analog to the textile industry. Several coils with linked micro parts are joined in an identified sequence. For enabling this synchronization an assessment of the occurring trends as well as the storage of the sorted parts is necessary. The identification of the joining sequence is performed through matching coils regarding their tolerances. Two major fields, a mathematical as well as a technological, are identified for investigation. An overview of the technological and mathematical process configuration as well as the consolidation of these approaches is presented in this paper.

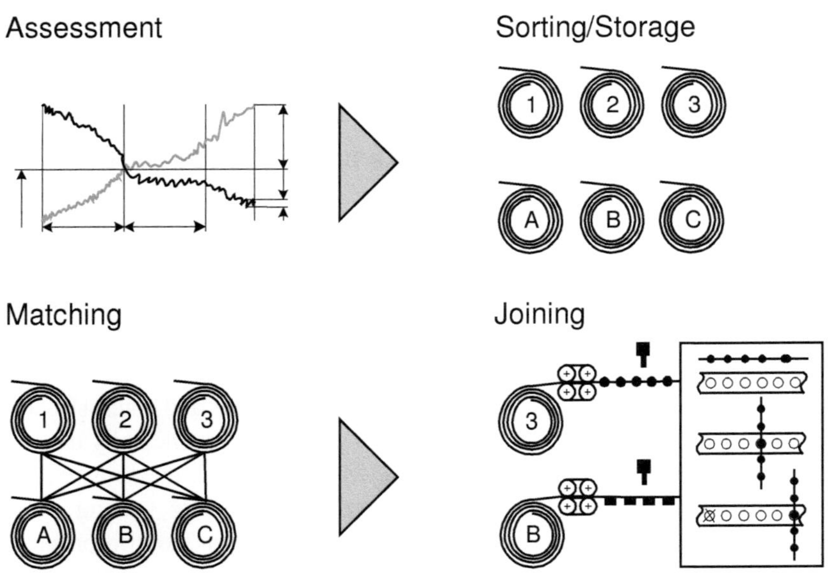

Fig. 1. Trend based joining of linked micro parts

2 Exemplification for Linked Micro Parts

The fit within assembled micro components depends on the geometric properties of the joining partner. In contrast to macro parts, micro parts like valves are commonly changed as modules. This requires an individual fit of these joining partners. It is not mandatory that the micro parts meet tolerances for enabling interchangeability. The linked production of these parts enables an identification of time depending changes of the geometry of the parts. Regarding an example of cups and spheres, the processes for implementing a trend depending joining is investigated. In this case the diameter is notably relevant regarding the tolerances for joining parts. It is assumed that the spheres are produced out of wire as line type in two steps. Within the first step a material accumulation through laser melting ([7]) and feeding is performed. Within the second step the material accumulation is formed through rotary swaging ([8]). The cups are produced out of belt material and are formed through a deep drawing process ([9]). The forming processes of these two kinds of linked parts are depicted in figure 2.

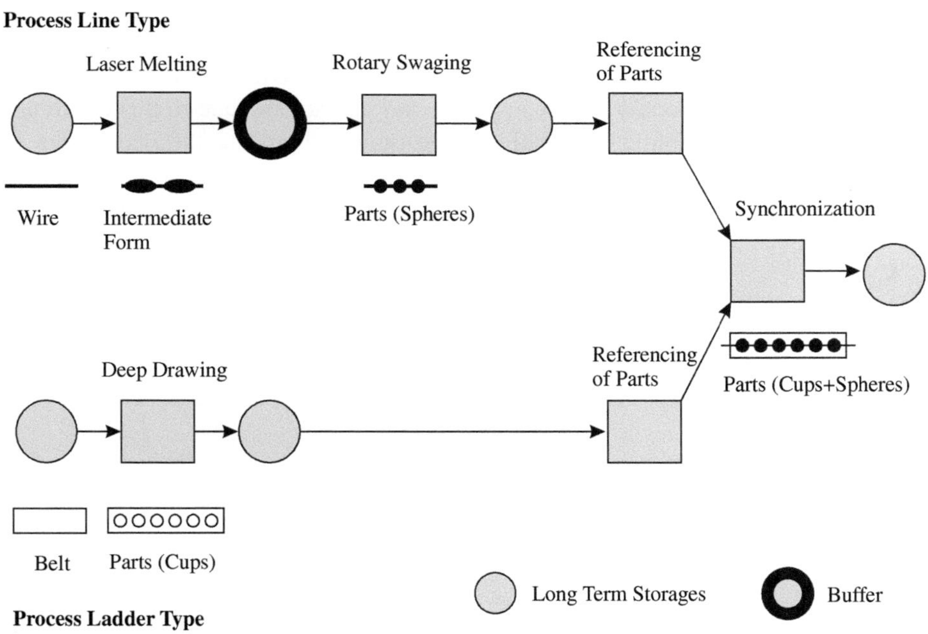

Fig. 2. Process chains of the introduced example

Besides the forming steps a referencing, storing and buffering of the parts is required. This is mandatory regarding the synchronization of the processes due to varying process times. A STEP based product model for micro formed linked parts has been developed in [10]. This enables a production specific data exchange through the whole process chain.

Considering the targeted trend based joining within the synchronization point, the assessment, sorting, storage, matching and joining of the spheres and cups has to be investigated. The mathematical approach for a widening of the tolerance field, that considers the appearing trends of the first batch as well as the technological approach are introduced in the following.

3 Tolerance Field Widening

Within the mathematical approach the investigation of a methodology for assessing and matching of the measured geometrical properties of the micro parts is addressed. As mentioned, the sequential production offers the usage of appearing trends. This is the basis for widening the tolerance field by matching sections of linked parts, which exhibit similar tolerance field trends. Thereby, the ascertainment of the trend of the first produced batch is utilized for a design improvement to optimize the output of well joined parts by adjusting geometric parameters like nominal values. Therefore, geometric properties like the diameter, which is considered in the example of cups and spheres must be measured. It is assumed that trends occur due to abrasive and adhesive wearing of the tools [8,9,11]. Wear is a process that depends on various factors. Due to this the trends could differ. For enabling a versatile usage of the approach, the occurrence of linear functions, step functions with changing trends as well as opposed trends are considered. Examples for assumed occurring trends are depicted in figure 3.

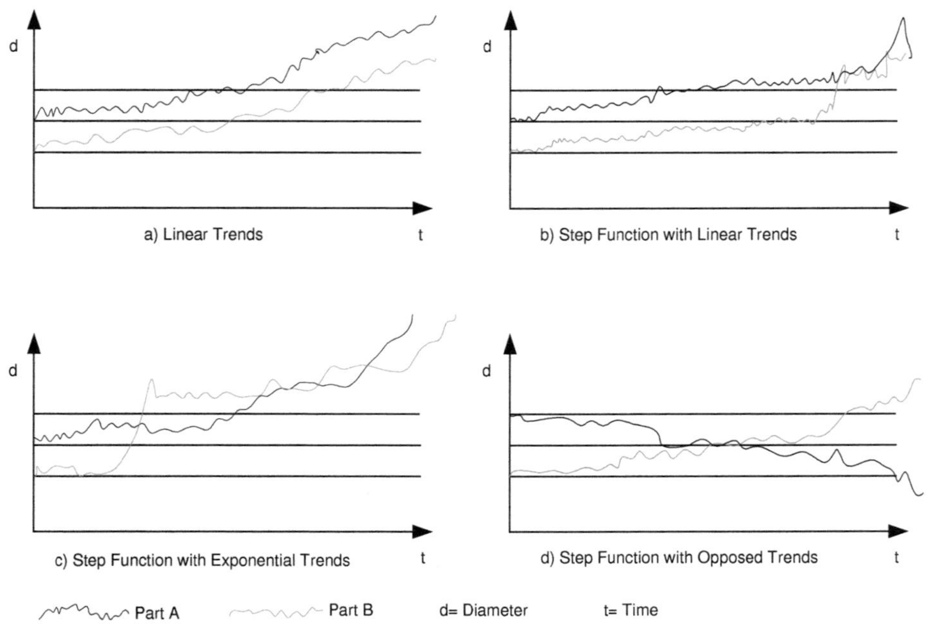

Fig. 3. Examples for occurring trends

This methodology is not used for a process adjustment . For adjusting the nominal value according to the first produced batch, a conjunction of a cluster algorithm, an optimization and a heuristic has to be investigated, as depicted in figure 4. Clustering is essential for a flexible and trend depending assessment of the measured diameters. Cluster algorithms for networks [12] could be used, because the interconnection of the parts must be considered. In this case the parts are vertices of the network and the interconnections are interpreted as edges.

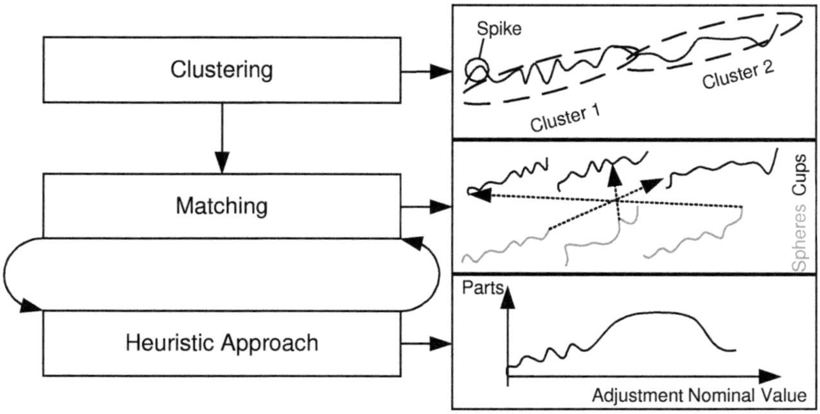

Fig. 4. Schema of the tolerance field widening

Basically, cluster algorithms for networks are building groups regarding vertex similarities or in accordance to fitness measures over a set of possible clusters. Regarding the vertex similarity, similarities of the connectivity as well as properties like values could be used for building groups with close-by values [12]. Considering the connectivity means that vertices belonging to the same cluster should be strongly connected [13]. Since the vertices are connected equally there is no possibility of dividing groups through connectivity measures. The initializing of groups must be done regarding trend specific similarities that have to be ascertained. According to the example of cups and spheres the diameter of the parts must be considered. Additionally, the clustering results are used as defined sections for the matching. Especially the identification of spikes would be a key issue. These parts must be discharged like elucidated in [5]. Otherwise, small sections must be joined and high handling efforts would be necessary. These parts must be marked and discharged afterwards. For enabling an application of clusters and identified spikes from the first batch to further batches, a referencing point would be necessary. In the first instance, it has to be investigated how the clustering results of these two different linked parts with different trends could be adapted to synchronization. A high number of different section sizes would increase the handling effort.

An optimization for the identification of the sequence of the composed sections that enables a maximized output is necessary in the second step of the methodology. Therefore, matching algorithms must be chosen and evaluated within the investigation of the methodology. Based on the result of the clustering a consideration as bipartite graphs is possible. According to [14], bipartite graphs are defined as graphs with two different kinds of vertices that are interconnected through edges starting in one group and ending in the other. As depicted in figure 5, the resulting clusters are interpreted as bipartite graphs. Since the sections are self-contained groups, it is possible to separate these groups for optimizing the sequence. For a maximized output of well joined parts it is necessary to change the sequence of the sections and join them regarding the tolerance fields. The consideration as bipartite graphs enables the usage of bipartite matching methods like [15], for improving the sequence. To begin with, it must be investigated how the values of the clusters could be used for representing the tolerance and the trend of the cluster.

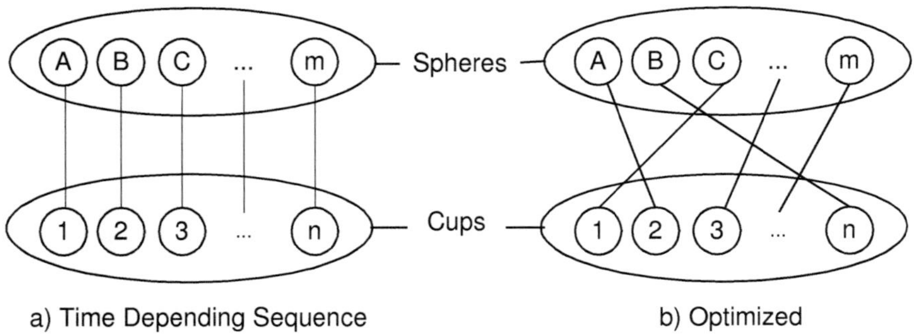

Fig.5. Resulting clusters of linked parts as bipartite Graph

Through an adjustment of the nominal value it is possible to enlarge the number of well joined parts. It is assumed that through the adjustment no major changes of the trend developments appear. Thereby, the adjustment leads to a shifting of the graph of the trend along the Y-axis and the diameter. The evaluation of the adjustment is done through another iteration of optimization with matching. Heuristic approaches like hill-climbing search ([16]) are used to decrease the running time, because computing of small adjustments of the nominal value is very time consuming. In the first instance, the identification and usage of sections with significant trends for the adjustment must be investigated.

Due to changing materials and material properties the production process could fluctuate [1]. Thereby it is possible, that the changeover between sections as well as trend developments could alternate. Therefore, especially at points where trends and sections change, a defined number of parts around the changeover should be measured carefully. For enabling an independent system without having subsequent required handling effort the usage of current process, trend and material data could be used for the implementation of a self-learning system as depicted in figure 6. This further advancement of the mathematical methodology enables a material depending adjustment of the diameter in accordance to present material characteristics without an assessment of one produced test batch. This advancement must be addressed subsequent to the widening of the tolerance field.

However, as a first step the tolerance field widening is investigated considering the first batch of micro formed parts. The explained conceptual approach has to be investigated and verified through tests within the CRC 747.

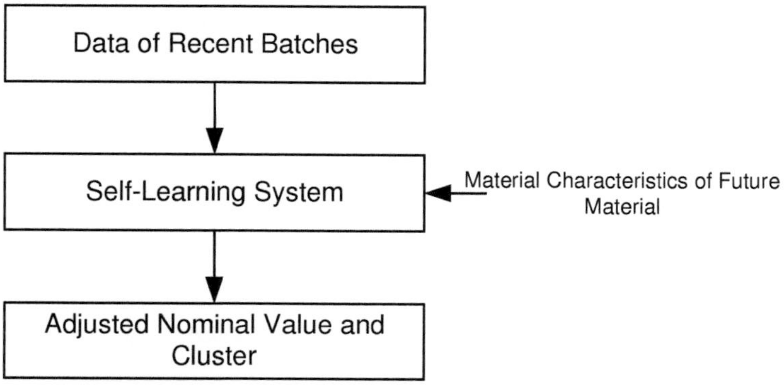

Fig.6. Schema of the material depending design modification

4 Synchronization Station

The technical approach of joining two different process chains of linked micro parts includes the constraints of the technological challenges, the requirements of the mathematical methodology as well as the process requirements for matching the linked parts. Technical solutions for matching ladder and line-type micro parts in the synchronization point already exist [18] and strategies for extraction of scrap parts from linked parts are addressed in [5], where the two concepts, cutting without loss of connection and with loss of connection are presented. Besides the extraction of scrap parts these methods are also required within the synchronization station regarding the tolerance field widening. Considering spikes it must be possible to discharge those. Therefore, the synchronization station must include production processes such as rotary swaging [19] for wire or belt joining.

However, the challenge of synchronizing processes with different cycle times and small tolerances addresses the necessity of a storage and buffer for linked micro parts as shown in figure 2. Both, buffer with high dynamics and low capacity as well as long term storages with low dynamics but high capacity are required and therefore developed and evaluated.

Due to the high capacity and process reliability, the well-known concept for spooling and unspooling wire from coils is adapted for the storage of micro components. Issues regarding damage free spooling of multiple layers are addressed by using protection material between each layer, which also must be inserted in synchronization with all processes. Tests regarding spooling and unspooling as well as usability and handling of layers are necessary for implementing these. Especially the protection of the layers is important regarding the spooling and unspooling as well as rewinding of linked parts. These processes could not be implemented if the parts were damaged. The possibility of rewinding is presupposed through the mathematical approach. This is required if opposing trends occur.

Furthermore, in cooperation with the mathematical methodology the number of used coils for storing the sequences, similar to the textile industry, must be investigated and defined. For a minimal effort the number should be low. On the other hand it has to be ensured that the required section is available for joining.

5 Summary and Outlook

Micro forming still challenges the industries due to occurring size effects, high handling efforts and tight tolerances. The production of linked parts facilitates an up scaling of the parts and eases the handling. Furthermore, the linked forming of micro parts enables an identification and utilization of the appearing trends. Within this first approach it has been introduced how the widening of the tolerance field enables an adjustment of the nominal value.

Two major fields of investigation for enabling a paradigm change in micro production are introduced in the paper. Within the conceptual mathematical approach of the paradigm change and the tolerance field widening, it has been shown how the measured trends could be processed and displayed as bipartite graphs. This enables the usage of matching algorithms. Furthermore, the necessity of using heuristic approaches is explained. Mathematical approaches for the identification of trends, spikes, applicable sizes of sections, matching and adjusting the nominal value have to be investigated regarding the tolerance field widening.

For implementing the synchronization as the second partition of the paradigm change, presupposes of the tolerance field widening and process interdependencies must be investigated. Especially the occurrence of different cycle times challenges the joining of the micro parts. Within the technological field of investigation, the interdependencies of the processes as well as the handling and allocation of the linked parts have to be investigated.

Acknowledgments The authors gratefully acknowledge the financial support by Deutsche Forschungsgemeinschaft (DFG, German Research Foundation) for Subproject C5 "Teileverbunde" within the SFB 747 (Collaborative Research Center) "Mikrokaltumformen - Prozesse, Charakterisierung, Optimierung".

References

1. Geiger M, Kleiner M, Eckstein R, Tiesler N, Engel U (2001) Microforming. Annals of the CIRP 50 (2):445-462

2. Vollertsen F (2008) Categories of size effects. Prod Eng Res Devel 2:377–383
3. Vollertsen F, Biermann D, Hansen HN, Jawahir IS, Kuzmann K (2009) Size effects in manufacturing of metallic components. CIRP Ann Manuf Technol 58(2):566–587
4. Hesselbach J, Wrege J, Raatz A (2006) Mikromontage. In: Lotter B, Wiendahl H-P (eds.) Montage in der industriellen Produktion. Springer Berlin, Heidelberg, New York: 463-482 (in German)
5. Tracht K, Weikert F (2013) Handling of micro parts. In: Vollertsen F (ed) Micro metal Forming. Springer, Heidelberg: 331-343
6. Tracht K, Schenck C, Weikert F, Kuhfuß B (2010) Conveyance of micro-cold-formed parts in a linkage. wt Werkstattstechnik Online 100(11/12): 864-868 (in German)
7. Vollertsen F, Sakkiettibutra J (2010) Different types to use laser as a forming tool. Physics Procedia 5: 193-203
8. Böhmermann F, Hasselbruch H, Herrmann M, Riemer O, Mehner A, Zoch H-W, Kuhfuss B (2015) Dry rotary swaging. wt Werkstattstechnik Online 105(11/12):830-835 (in German)
9. Flosky H, Vollertsen F (2014) Wear behaviour in a combined micro blanking and deep drawing process. CIRP Annals - Manufacturing Technology 63:281-284
10. Weikert F, Tracht K (2016) STEP product model for micro formed linked parts.
11. Boher C, Attaf D, Penazzi L, Levaillant C (2005)Wear behaviour of the radius portion of a die in deep-drawing: Identification, localisation and evolution of the surface damage. Wear 25:1097-1108
12. Schaeffer S E (2008) Graph clustering. Computer Science Review 1(1):27-64
13. Hartuv E, Shamir R (2000) A clustering algorithm based on graph connectivity. Information processing letters 76 (4-6):175-181
14. Diestel R (2005) Graph Theory. Graduate Texts in Mathematics 173, 3rd edition
15. Munkres J (1957) Algorithms for the Assignment and Transportation Problems. Journal of Society for Industrial and Applied Mathematics 5(1):32-38
16. Hernandez D, Gras R, Appel R (2008) Neighborhood functions and hill-climbing strategies dedicated to the generalized ungapped local multiple alignment. European journal of operational research 185: 1276-1284
17. Werner J. (2009) Methoden zur roboterbasierten förderbandsynchronen Fließmontage am Beispiel der Automobilindustrie. Herbert Utz Verlag, München (in German)
18. Moumi E, Wilhelmi P, Kuhfuss B, Schenck C, Tracht K (2014) Wire joining by rotary swaging. Procedia Engineering 81:2012-2017

Combined Structural-Dimensional Synthesis of Robot Manipulators for Minimal Energy Consumption

Daniel Ramirez[12], Jens Kotlarski[1], and Tobias Ortmaier[1]

[1] Institute of Mechatronic Systems, Leibniz Universität Hannover, Appelstr. 11a, 30167 Hannover, Germany
{daniel.ramirez, jens.kotlarski, tobias.ortmaier}@imes.uni-hannover.de,
[2] Mechanical Engineering Applications and Research Group (GEAMEC), Universidad Santo Tomás, Cra 9 No. 51-11, Bogotá, Colombia

Abstract. The present work addresses the synthesis of robot manipulators for energy consumption minimization considering the influence of the kinematic architecture. The introduced approach aims to find a manipulator with the lowest energy consumption while – unlike related publications – taking all task-suitable architectures into account. The proposed method automatically generates a set of task-suitable architectures together with their specific adjustable design parameters. Inverse dynamics is calculated in symbolic form for each architecture in order to reduce the computational effort during the synthesis. Furthermore, the optimal manipulator for an exemplary pick and place operation is determined by a particle swarm optimization, using the energy consumption as optimization index.

Keywords: Structural synthesis, Dimensional synthesis, Energy consumption, Robot modeling

1 Introduction

Due to the increasing significance of the energy efficiency, specially in the industry, energy consumption is one of the main points taken into account for the performance evaluation of a robot. The energy efficiency of a robot can be improved in different ways, e.g. through special trajectory generation methods [1,2] or through energy exchange between the axes [3]. However, the whole geometry of a manipulator can also be optimized with respect to energetic aspects. In order to obtain the lightest possible design, the optimal dimensions, motors, and gearboxes of an anthropomorphic arm were determined in [4]. In [5], the geometry of a robotic arm was optimized taking workspace, stiffness, and the weight of the links into account. Multi-objective optimization was also used in [6] to maximize the manipulability and minimize at the same time the torque required to perform a trajectory. In this case, the cross-section was also considered during the optimization.

However, all studies consider only one specific kinematic design ignoring other possible architectures that can be obtained from the structural synthesis [7, 8]. Thus, the manipulator obtained from the geometrical optimization is not necessarily the most efficient one for the given task.

The present work introduces a combined structural and dimensional synthesis of serial robot manipulators for reduction of the energy consumption. The proposed method is able to synthesize manipulators with up to six degrees of freedom (DOF) while taking prismatic (P) as well as revolute (R) joints into account. In the first step, all task-suitable architectures are determined and their geometric optimization parameters are extracted (see Sect. 2). Afterward, the numerical solution of the inverse kinematics as well as the dynamics modeling for a generic serial manipulator are introduced in Sect. 3. The fourth section addresses the definition of the optimization problem for highest energy efficiency. In the last section, the capability of the approach is demonstrated by synthesizing a manipulator for an exemplary pick and place task with four DOF.

2 Generation of Suitable Architectures and Extraction of their Optimization Parameters

In the first step of the combined synthesis, all architectures are identified, that are able to perform the desired task. Furthermore, the geometric parameters of each suitable architecture, which can be used as optimization parameters without changing the end effector degrees of freedom (EE-DOF), are extracted. This allows for a reduced computational effort during the optimization process, as well as a subsequent performance analysis of different architectures.

The necessary DOF to execute a given task is defined using the required motion vector $\boldsymbol{\xi}_{\text{req}}$:

$$\boldsymbol{\xi}_{\text{req}} = \left[\xi_{\text{req}_1}, \xi_{\text{req}_2}, \xi_{\text{req}_3}, \xi_{\text{req}_4}, \xi_{\text{req}_5}, \xi_{\text{req}_6}\right]^{\text{T}} , \tag{1}$$

where ξ_{req_1}, ξ_{req_2}, and ξ_{req_3} represent the translational DOF in direction of the axes x, y, and z while ξ_{req_4}, ξ_{req_5}, and ξ_{req_6} represent the rotational DOF around the axes x, y, and z, respectively. It is worth noticing that $\boldsymbol{\xi}_{\text{req}}$ corresponds to the EE velocity of a serial manipulator $\boldsymbol{\xi}$:

$$\boldsymbol{\xi} = \left[\xi_1, \xi_2, \xi_3, \xi_4, \xi_5, \xi_6\right]^{\text{T}} = \begin{bmatrix} \boldsymbol{v}_n \\ \boldsymbol{\omega}_n \end{bmatrix} = \boldsymbol{J}\dot{\boldsymbol{q}} , \tag{2}$$

with $\boldsymbol{J}$ being the Jacobian matrix of the manipulator, $\boldsymbol{v}_n = \left[\xi_1, \xi_2, \xi_3\right]^{\text{T}}$ the EE linear velocity, $\boldsymbol{\omega}_n = \left[\xi_4, \xi_5, \xi_6\right]^{\text{T}}$ the EE angular velocity, and $\dot{\boldsymbol{q}}$ the joint rates vector.

The architecture of a serial manipulator can be described through a symbolic matrix of Denavit-Hartenberg (DH) parameters. Some of these parameters have to remain constant in order to preserve the EE-DOF, but the residual can be used as optimization parameters.

The suitability of a manipulator with respect to the task is evaluated using three conditions:

1. the null entries of $\boldsymbol{\xi}_{\text{req}}$ and $\boldsymbol{\xi}$ have to be the same. It means, if $\boldsymbol{\xi}_{\text{req}_i} = 0$ for any $i = 1 \ldots 6$, then $\boldsymbol{\xi}_i$ should also be zero for the same i and for any $\dot{\boldsymbol{q}}$,
2. the rank of the Jacobian (rank($\boldsymbol{J}$)) must be equal to the total number of required DOF,
3. the independence of each EE DOF with respect to each DH parameter has to be verified in order to eliminate isomorphisms.

After applying this conditions to the generated set of manipulators, it is possible to obtain a minimal set of architectures (i. e. without isomorphisms) to be considered during the subsequent optimization [9].

To illustrate the representation of different architectures in the using the proposed notation, the DH parameters of two solutions from the minimal set of suitable architectures for a task requiring three translational DOF and one rotational DOF around the z-axis are shown in Table 1. Both solutions correspond to the joint sequence RPRP, but entail different orientation of the joint axes (see Fig. 1). Therefore, they are identified as RPRP_1 and RPRP_2. In the latter case, the optimization parameters vector results to

$$\boldsymbol{p} = [\theta_4,\, d_1,\, d_3,\, a_1,\, a_2,\, a_3,\, a_4,\, \alpha_4] \; . \tag{3}$$

The remaining DH parameters have fixed values namely $\theta_2 = 0$, $\alpha_1 = \pi/2$, $\alpha_2 = \pi/2$, and $\alpha_3 = 0$.

Table 1. DH parameters of the architectures RPRP_1 (a) and RPRP_2 (b)

(a) RPRP_1

R/P	θ_i	d_i	a_i	α_i
R	q_1	d_1	a_1	0
P	θ_2	q_2	a_2	0
R	q_3	d_3	a_3	$\pi/2$
P	θ_4	q_4	a_4	α_4

(b) RPRP_2

R/P	θ_i	d_i	a_i	α_i
R	q_1	d_1	a_1	$\pi/2$
P	0	q_2	a_2	$\pi/2$
R	q_3	d_3	a_3	0
P	θ_4	q_4	a_4	α_4

3 Kinematic and Dynamic Modeling

In the combined structural-dimensional robot synthesis, the inverse kinematics is solved using a numerical approach, which exploits the advantages of both the pseudoinverse and the transpose of the Jacobian matrix.

The first case corresponds to the Newton-Raphson method for solving nonlinear kinematic equations [12]. Though the approach usually has a high convergence rate, it diverges when the Jacobian $\boldsymbol{J}$ is singular. The second case corresponds to the gradient method for the solution of a system of nonlinear equations [13], utilizing the transpose of $\boldsymbol{J}$ instead of the pseudoinverse. Since the inversion of the Jacobian is not necessary in this case, the approach is not affected by singularities. However, the convergence is slower in comparison to the Newton-Raphson method.

The proposed approach switches between the pseudoinverse and the transpose of $\boldsymbol{J}$ depending on the condition number κ of the Jacobian $\boldsymbol{J}$. In this way, the value of κ is employed to determine the proximity to a robot singular configuration. Given a threshold $\kappa_{\max}$ the pseudoinverse is used if $\kappa < \kappa_{\max}$ and the transpose if $\kappa \geq \kappa_{\max}$ [10].

The dynamics of a general serial manipulator with n joints can be described using the Lagrange formulation [13]:

$$\frac{\mathrm{d}}{\mathrm{d}t}\left(\frac{\partial L}{\partial \dot{\boldsymbol{q}}}\right) - \frac{\partial L}{\partial \boldsymbol{q}} = \boldsymbol{\tau}_{\mathrm{L}}\,, \tag{4}$$

where $\boldsymbol{\tau}_{\mathrm{L}} \in \mathbb{R}^n$ corresponds to the vector of the external torques/forces and L the Lagrangian of the robot. If the links of the manipulator are considered as rigid bodies and the elasticity in the actuators and gearboxes is neglected, equation (4) yields the equation of the inverse dynamics:

$$\boldsymbol{M}\,(\boldsymbol{q})\,\ddot{\boldsymbol{q}} + \boldsymbol{c}\,(\boldsymbol{q},\,\dot{\boldsymbol{q}}) + \boldsymbol{g}\,(\boldsymbol{q}) = \boldsymbol{\tau}_{\mathrm{L}}\,, \tag{5}$$

where $\boldsymbol{M}\,(\boldsymbol{q}) \in \mathbb{R}^{n\times n}$ is the mass matrix, $\boldsymbol{c}\,(\boldsymbol{q},\,\dot{\boldsymbol{q}}) \in \mathbb{R}^n$ the Coriolis and centrifugal effects vector, and $\boldsymbol{g}\,(\boldsymbol{q}) \in \mathbb{R}^n$ is the gravitational effect vector. The external torques/forces in (5) correspond to the torques/forces of the actuators, since friction effects are neglected in the conceptual design stage.

4 Optimization Strategy

In order to find the configuration (architecture and dimensions) with the minimal energy consumption for a given task, the total energy consumed to perform the task E_{Total} is utilized as cost function to optimize every suitable architecture from Sect. 2. At the beginning of the optimization process, equation (5) is calculated in symbolic form and the corresponding MATLAB-functions are automatically generated for each architecture. In contrast to iterative methods, the use of exact solutions of the dynamics allows for a significantly reduced simulation time.

As a result of the inverse dynamics, the torque/force $\tau_{\mathrm{L},i}$ of the i-th actuator can be employed to determine the corresponding motor torque τ_i using the combined inertia (motor and gearbox) $J_{\mathrm{g},i}$ and the gearbox ratio ρ_i:

$$\tau_i\,(t) = \frac{\tau_{\mathrm{L},i}\,(t)}{\rho_i} + J_{\mathrm{g},i}\,\rho_i\,\ddot{q}_i(t)\,. \tag{6}$$

The power consumption of each motor P_i is the product of the motor torque/force and the motor speed:

$$P_i\,(t) = \tau_i\,(t)\;\rho_i\,\dot{q}_i(t)\,. \tag{7}$$

A common practice, especially in industrial robots, is to connect the drives of the robot over a common DC-bus in order to allow electrical power exchange [1]. Hence, the resulting total power P_{sum} is calculated:

$$P_{\text{sum}}(t) = \sum_{i=1}^{n} P_i(t) \,. \tag{8}$$

Since only positive values of P_{sum} have to be considered to determine the energy consumption [3], a heaviside function Θ is used to set all negative values of P_{sum} to zero. The energy $E_{j_{\text{traj}}}$ consumed during the execution of a specific trajectory is obtained by integration over the trajectory time $\Delta t = t_1 - t_0$:

$$E_{j_{\text{traj}}} = \int_{t_0}^{t_1} \Theta\, P_{\text{sum}}(t)\, \mathrm{d}t \,. \tag{9}$$

If the task comprises several trajectories, the total energy E_{Total} results from the sum of the energy of each trajectory:

$$E_{\text{Total}} = \sum_{j_{\text{traj}}=1}^{N_{\text{traj}}} E_{j_{\text{traj}}} \,. \tag{10}$$

The optimization aims to find the optimal geometric parameters vector $\boldsymbol{p}^*$ that minimizes E_{Total}:

$$\boldsymbol{p}^* = \arg\min_{\boldsymbol{p}} \left(E_{\text{Total}}(\boldsymbol{p}) \right) , \tag{11}$$

subject to

$$\begin{aligned} \tau_{i,\text{max}} &\leq \tau_{\text{peak,motor}_i} \,, \\ \tau_{i,\text{rms}} &\leq \tau_{\text{nom,motor}_i} \,, \\ \rho_i\, \dot{q}_{i,\text{max}} &\leq \omega_{\text{peak,motor}_i} \,, \\ \boldsymbol{p}_{\text{min}} &\leq \boldsymbol{p} \leq \boldsymbol{p}_{\text{max}} \,, \end{aligned} \tag{12}$$

with $\tau_{\text{peak,motor}_i}$, $\tau_{\text{nom,motor}_i}$, $\omega_{\text{peak,motor}_i}$ being the maximum permissible torque, nominal torque, and maximum speed of the i-th joint motor respectively. $\tau_{i,\text{max}}$ and $\tau_{i,\text{rms}}$ represent the maximum value and rms value of the torque exerted by the motor in the the i-th joint. $\dot{q}_{i,\text{max}}$ is the maximum speed of the i-th joint. Furthermore, the lower and upper bounds of $\boldsymbol{p}$ are $\boldsymbol{p}_{\text{min}}$ and $\boldsymbol{p}_{\text{max}}$.

The particle swarm optimization algorithm (PSO) is chosen to optimize each suitable architecture with respect to (11).

5 Exemplary Results

The capability of the proposed method is demostrated on an exemplary pick and place task. The task consists of moving a 10 kg mass from the start pose

$\boldsymbol{x}_{\mathrm{EE},0}$ through the poses $\boldsymbol{x}_{\mathrm{EE},1}$, $\boldsymbol{x}_{\mathrm{EE},2}$, and $\boldsymbol{x}_{\mathrm{EE},3}$, which are listed in Table 2. The required poses are described using the position and orientation of the end effector coordinate system $(\mathrm{CS})_{\mathrm{EE}}$ with respect to the robot base coordinate system $(\mathrm{CS})_0$, whose position and orientation remain constant during the optimization. The coordinate φ in Table 2 corresponds to the rotation around a vertical axis (parallel to z-axis). The required motion vector for this task is

$$\boldsymbol{\xi}_{\mathrm{req}} = \left[\xi_{\mathrm{req}_1}, \xi_{\mathrm{req}_2}, \xi_{\mathrm{req}_3}, 0, 0, \xi_{\mathrm{req}_6}\right]^{\mathrm{T}} . \tag{13}$$

All architectures that fulfill (13) are shown in Fig. 1.

Table 2. Required poses of the task

Pose	x [m]	y [m]	z [m]	φ [rad]
$\boldsymbol{x}_{\mathrm{EE},0}$	−0.5	0.5	0	$-\pi/2$
$\boldsymbol{x}_{\mathrm{EE},1}$	0.5	0.8	0.7	0
$\boldsymbol{x}_{\mathrm{EE},2}$	0.5	0.3	0.7	0
$\boldsymbol{x}_{\mathrm{EE},3}$	0.5	0.2	0.1	$-\pi/2$

In order to ensure the task feasibility, a point to point (PTP) interpolation is calculated in the joint space using a polynomial function of degree five [15]. The total energy consumption of the full task is calculated using (10).

Since the synthesis process is part of the conceptual phase of a robot design, some assumptions have to be taken for estimating physical properties of the robot links and actuators in the inverse dynamics calculation. In this example, all robot links are assumed to be constructed using standard aluminum profiles (80×80 light [16]). For simplicity, two types of motors and gearboxes are used, one for the first and second joint, and one for the third and fourth joint. In the case of prismatic joints, the motors are connected to a spindle instead to a gearbox

The resulting energy consumption of the optimized architectures is shown in Fig. 2. It is important to note, although the PPRP architecture is able to perform the task in general, a solution that fulfill the constraints 12 does not exist. As can be observed, the best performance belongs to a manipulator with an RRRP architecture. Its optimized DH parameters are presented in Table 3. Although several simplification were used specially in the dynamic modeling, the difference between the energy consumption of the architectures is considerable. Compared with the architecture with the lowest energy demand (RRRP), the architecture with the highest consumption (PRRR) presents a difference of 461 %. Furthermore, the difference between the two architectures with the best performance (RRRP, and RPRP_2) is 10 %.

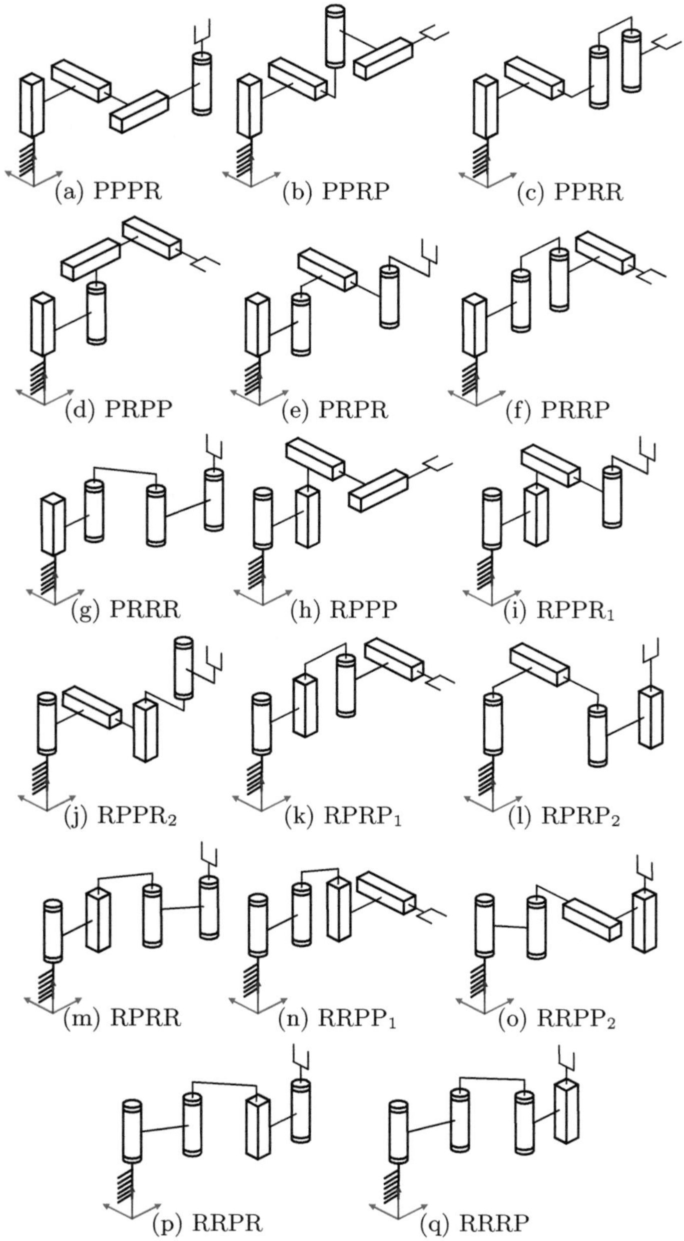

Fig. 1. Architectures generated for $\boldsymbol{\xi}_{\text{req}} = [\xi_{\text{req}_1}, \xi_{\text{req}_2}, \xi_{\text{req}_3}, 0, 0, \xi_{\text{req}_6}]^{\text{T}}$

6 Conclusion

This paper presents an approach to find an energy efficient serial robot configuration using a task-related combined structural and dimensional synthesis.

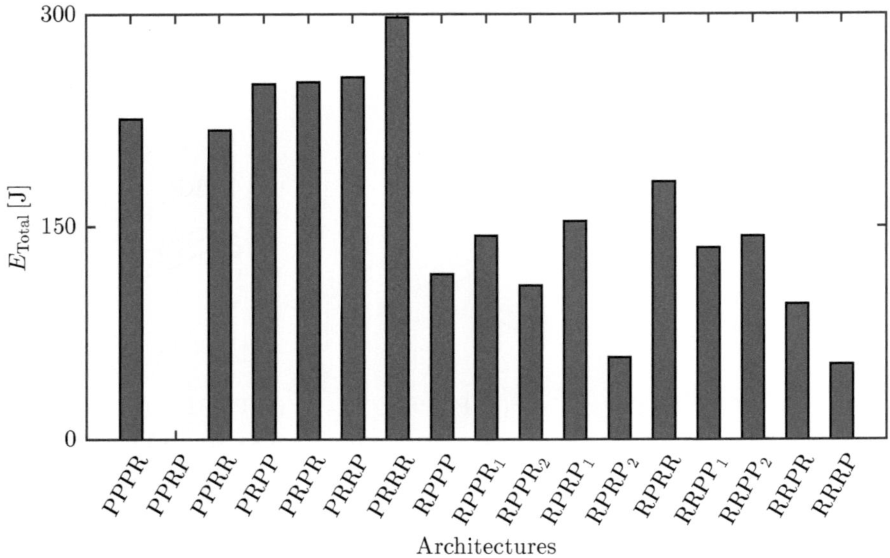

Fig. 2. Energy consumption for every architecture of Fig. 1 after the optimization procedure

Table 3. DH parameters of the manipulator with the lowest energy consumption

R/P	θ_i [rad]	d_i [m]	a_i [m]	α_i [rad]
R	q_1	0.172	0.839	0
R	q_2	0.174	0.821	0
R	q_3	0.197	0.343	0
P	1.141	q_4	0.170	2.960

A set of suitable architectures is automatically generated and their geometric optimization parameters are extracted. The inverse dynamics is calculated in symbolic form for each architecture and utilized to calculate the specific energy consumption. This measure is used as cost function to optimize the dimensions of each architecture. Afterward, the manipulator with the lowest energy consumption is chosen as the optimum for the given task. The advantages of considering all task-suitable architectures are presented for an exemplary pick and place operation.

Besides the determination of the optimal combination of architecture and dimensions, the approach allows for a performance comparison between several architectures. Since the method is automatically performed, it significantly reduces the effort for the conceptual design of new manipulators. Future work addresses additional features, e.g. avoidance of obstacles, or automatic selection of the actuators.

References

1. Hansen C., Kotlarski J., Ortmaier T.: Enhanced approach for energy-efficient trajectory generation of industrial robots In: Proceedings of the 2012 IEEE International Conference on Automation Science and Engineering, Korea, 2012
2. Saravanan R., Ramabalan S., Balamurugan C.: Evolutionary optimal trajectory planning for industrial robot with payload constraints. The International Journal of Advanced Manufacturing Technology. **38**(11), 1213–1226 (2007)
3. Hansen C., Kotlarski J., Ortmaier T.: Path planning approach for the amplification of electrical energy exchange in multi axis robotic systems. In: Proceedings of the 2013 IEEE International Conference on Mechatronics and Automation, Japan, 2013
4. Zhou L., Bai S., Hansen M.: Integrated dimensional and drive-train design optimization of a light-weight anthropomorphic arm. Robotics and Autonomous Systems. **60**(1) 113–122 (2012)
5. Castejon C., Carbonare G., Garcia J., Ceccarelli M.: A multi-objective optimization for designing service robots. In: Proceedings of The 12th IFToMM World Congress, France, 2007.
6. Saravanan R., Ramabalan S., Ebenezer N., Natarajan R.: Evolutionary bi-criteria optimum design of robots based on task specifications. The International Journal of Advanced Manufacturing Technology. **41**(3) 386–406 (2008)
7. Caro S., Khan W., Pasini D., Angeles J.: The rule-based conceptual design of the architecture of serial schoenflies-motion generators. Mechanism and Machine Theory. **45**(2) 251–260 (2010)
8. Kuo C., Dai J.: Structural synthesis of serial robotic manipulators subject to specific motion constraints. In: Proceedings of the ASME 2010 International Design Engineering Technical Conferences & Computers and Information in Engineering Conference, Montreal, 2010
9. Ramirez D., Kotlarski J., Ortmaier T.: Automatic generation of a minimal set of serial mechanisms for a combined structural - geometrical synthesis. In: Proceedings of the 14th IFToMM World Congress, Taipei, 2015
10. Ramirez D., Kotlarski J., Ortmaier T.: Combined Structural and Dimensional Synthesis of Serial Robot Manipulators. In: Parenti-Castelli, V., Schiehlen, W. (eds) ROMANSY 21 - Robot Design, Dynamics and Control: Proceedings of the 21st CISM-IFToMM Symposium, June 20-23, Udine, Italy. Springer International Publishing (2016). doi:10.1007978-3-319-33714-2_23
11. Corke P.: Robotics, Vision and Control. Springer, 2011.
12. Schwarz H., Köckler N.: Numerische Mathematik. Vieweg Teubner Verlag, 2011
13. Sciavicco L., Siciliano B.: Modelling and Control of Robot Manipulators, 2nd ed. Springer-Verlag London, 2000
14. Thanh T., Kotlarski J., Heimann B., Ortmaier T.: A new program to automatically generate the kinematic and dynamic equations of general parallel robots in symbolic form. In: Proceedings of the 1st IFToMM International Symposium on Robotics and Mechatronics. **1** (2009)
15. Biagiotti L., Melchiorri C.: Trajectory Planning for Automatic Machines and Robots. Springer Berlin Heidelberg, 2008
16. smt montagetechnik [Online]. Available: http://www.alupark.com/alutechnik-545.pdf (verified Dec. 2016)
17. Lenze [Online]. Available: http://www.lenze.com/produkte/motoren/servomotoren/servo-synchronmotoren-mcs/ (verified Dec. 2016)

Development of a modular framework for contact force models

Tobias Osterloh and Jürgen Roßmann

Institute for Man-Machine-Interaction, RWTH Aachen, Ahornstraße 55, 52074 Aachen
osterloh@mmi.rwth-aachen.de, rossmann@mmi.rwth-aachen.de
WWW home page: http://www.mmi.rwth-aachen.de/

Abstract. Nowadays, various models of contact forces are widespread in physics-based simulators. However, these models are specific to particular applications. In order to broaden these contact models to a wide range of applications, and thus foster multi-purpose simulation capabilities, a more general treatment of contact forces is required. Based on the constraint force mixing principle a mathematical basis for a modular framework can be found, allowing to include continuous as well as discrete contact force models to dynamic simulations. This modularity gives a high degree of flexibility in contact simulation, delivering the possibility to configure the contact simulation to various application scenarios.

Keywords: Contact Simulation, Contact Force Models, Multibody Dynamics Simulation, Constraint Force Mixing

1 Introduction

Accurate simulation of contact forces is a very complex and an active research field in dynamics simulation. Contact simulation plays a key role in various industrial and robotic applications, such as grasping, inserting and assembly in peg-in-hole scenarios or in-orbit docking of satellites. As a multibody dynamics simulation is one of the key components of Virtual Testbeds [1] in the eRobotics approach [2] a highly precise contact simulation is required in order to reliably predict the interactive behavior of a real system.

There are various aspects that influence the accuracy of simulated contact models. First of all, the contact simulation is characterized by the mathematical formulation of the multibody dynamic simulation like Penalty based, Linear Complementarity Problem (LCP) or Quadric Programming approaches [3][4]. Furthermore, the fidelity of contact simulation highly depends on the utilized collision detection scheme and the accurate determination of the correct time of impact. For contact resolution, various discrete and continuous contact force models exist [5]. Furthermore, diverse friction models allow the simulation of phenomena such as microslip [6]. Another aspect that should be consider is the integration scheme used to determine the correct movements of the colliding bodies.

Regarding all these aspects and the various applications (see section 5) for a contact simulation, it becomes obvious that current physics-based simulations cannot succeed with the same accuracy and performance in various use case. This raises the need for a modular framework that facilitates the configuration of a contact simulation to the particular application. This modularization implies the possibility to choose between various contact force profiles, as shown in figure 1.

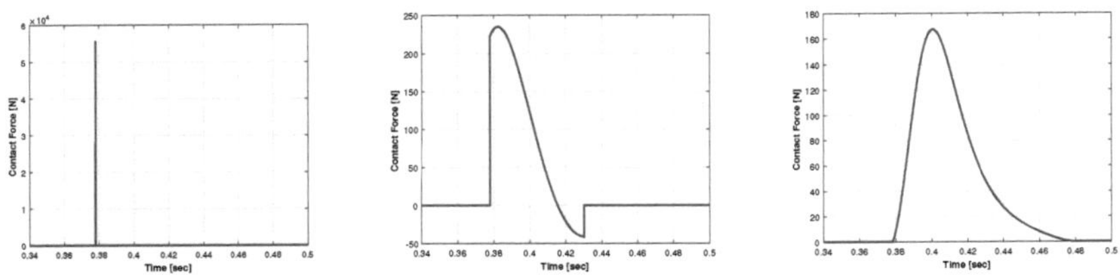

(a) Discrete contact force profile

(b) Spring damper contact force profile

(c) Non-linear spring damper contact force profile

Fig. 1: Various contact force profiles of a modular contact force framework

2 Related Work

In literature a variety of contact force models is known. The various models can be divided by discrete and continuous contact force models [7]. Discrete models describe the movement of the two colliding bodies before and after the collision with a single coefficient of restitution. In Newton's contact law the coefficient of restitution is given by the relative velocity of colliding bodies before and after the impact, whereas Poisson's contact law uses an impulse based calculation for the coefficient of restitution and Strong's hypothesis is an energy based formulation of the coefficient of restitution [7]. Despite the various formulations of the coefficient of restitution, every discrete contact model is able to determine the relative velocity after the collision [8]. One characteristic of discrete models is that they resolve the collision situation within a single integration time step. Discrete models are often used in impulse based dynamic simulations, requiring - especially in multi-contact situations - appropriate iteration techniques to establish a valid system state without penetration [9].

Due to its simplicity, an often used approach for contact resolution is a compliant spring damper model. It is straightforward to implement and gives good results for many use cases [10]. However, this contact force model introduces discontinuities to the contact force profile and may lead to negative contact forces at the end of the restitution phase, attracting the two colliding bodies, which is physically impossible behavior [11]. Reference experiments have shown that the contact force profile has a smooth transition from compression to restitution phase, where only positive contact forces occur [12]. To include these contact forces in dynamics simulation, primarily external forces are applied to the colliding bodies. Constraint force mixing can be used to include contact situations as constraints in a LCP based dynamics simulation. One advantage of constraint force mixing is a more stable and robust formulation and the possibility to resolve various contacts solving a single equation system, not needing special iteration techniques [10].

To avoid discontinuities in the contact force profile and negative contact forces, non-linear spring damper contact force models are used. There are plenty of parametrizations for this type of models [5]. All parametrizations were developed for special contact situations and various use cases. A typical approach to include these non-linear contact force models to a dynamic simulation is by applying external forces to the colliding bodies [11].

3 State of the Art

Multibody dynamics simulation in maximal coordinates is based on Newton-Euler equations. The acceleration of the rigid bodies in a multibody system is given by external forces $\vec{f}_{ext}$ applied to the bodies and the constraint forces $\vec{\lambda}$ caused by kinematic constraints like joints. Applying a simple Euler integration step with step size h leads to the velocity update.

$$\vec{v}(t+h) = \vec{v}(t) + \mathbf{M}^{-1}\left[\vec{f}_{ext} \cdot h + \mathbf{J}^T \cdot \vec{\lambda} \cdot h\right] \tag{1}$$

The constraint forces $\vec{\lambda}$ are calculated such, that the constraints will be fulfilled in the time step $t+h$. The vector $\vec{b}$ can be used to inject desired relative velocities to the equation, as it is the case for simulation of motors [13].

$$\mathbf{J}\mathbf{M}^{-1}\mathbf{J}^T \cdot h \cdot \vec{\lambda} + \mathbf{J}\left[\vec{v}(t) + \mathbf{M}^{-1} \cdot h \cdot \vec{f}_{ext}\right] = \vec{b} + \vec{a} \tag{2}$$

When it comes to contact simulation, the bilateral constraint has to be replaced by a unilateral constraint, allowing only to apply positive contact forces that prevent the colliding bodies form penetrating. This requirement can be formulated as complementarity constraint. For a contact situation two possibilities can occur. In the first case, two bodies collide leading to a positive contact force λ and a relative velocity $a = 0$, indicating a resting contact. In the second case, two colliding bodies have a positive relative velocity a, indicating a release of the contact, so the contact force is determined $\lambda = 0$. To correct penetrations for resting contacts, the factor $\vec{b}$ is used with Baumgarte stabilization [14].

$$\vec{a} \geqslant \vec{0}, \quad \vec{\lambda} \geqslant \vec{0}, \quad \vec{a}^T\vec{\lambda} = 0 \tag{3}$$

Equation (3) is the classical non-penetrating constraint. However, in order to simulate springiness of contacts, the constraint can be modified according to a spring damper contact model. In this case, the contact force is calculated based on the classical spring-damper model.

$$F_c = k \cdot x(t) + d \cdot v(t) \tag{4}$$

where $x(t)$ is the penetration depth and $v(t)$ is the relative velocity of the colliding bodies.

To consider this spring damper behavior in the equation system of the multibody dynamic simulation, equation (2) can be extended by the so called Constraint Force Mixing principle using the constraint force mixing coefficient c_{cfm}, the error reduction parameter k_{er} and the current deflection e of the contact [10] [15].

$$c_{cfm} = \frac{1}{hk + d} \qquad k_{er} = \frac{hk}{hk + d} \qquad e = -x(t) \tag{5}$$

In general, the constraint force mixing principle can be described with diagonal matrices. These matrices are used to modify the calculation of the constraint forces λ, introducing a spring damper behavior in the respective direction of the constraint given by the Jacobian $\mathbf{J}$.

$$\left[\mathbf{J}\mathbf{M}^{-1}\mathbf{J}^T + \frac{1}{h}\mathbf{C}_{cfm}\right] \cdot h \cdot \vec{\lambda} + \mathbf{J}\left[\vec{v}(t) + \mathbf{M}^{-1} \cdot h \cdot \vec{f}_{ext}\right] = \vec{b} + \frac{1}{h}\mathbf{K}_{er}\vec{e} \tag{6}$$

This generalized approach allows to robustly and efficiently treat multiple contact points in a single equation system, not needing any iteration techniques.

4 Modular Framework

Starting point for the modular framework is a general formulation of a non-linear spring damper system [11].

$$F_c = k \cdot x^n(t) + d \cdot x^p(t) \cdot v(t) \tag{7}$$

In literature a hysteresis damping factor $\chi = d \cdot x^p(t)$ is defined commonly. The damping factor χ is calculated depending on the current penetration $x(t)$, the initial relative contact velocity v_0 and mass properties of the colliding bodies. This allows to parametrize the damping of the non-linear contact force model to a particular contact situation [5].

Equation 7 can be linearized around the current point of penetration x_c and current relative velocity v_c with first order Taylor approximation. This linearization is performed every simulation step, describing an approximated linear behavior in a small area around (x_c, v_c), see figure 2.

$$F_c \approx F_c(x_c, v_c) + \left.\frac{\partial F_c}{\partial v}\right|_{(x_c,v_c)} \cdot [v(t) - v_c] + \left.\frac{\partial F_c}{\partial x}\right|_{(x_c,v_c)} \cdot [x(t) - x_c] \tag{8}$$

$$= C_1 \cdot v(t) + C_2 \cdot x(t) + C_3 \tag{9}$$

where C_1 and C_2 describe the vectors spanning the tangential plane of the linearization and C_3 the zero-crossing of the tangential plane. The stationary state of this linearized differential equation is $x_{stat} = \lim_{t \to \infty} x(t) = -\frac{C_3}{C_2}$.

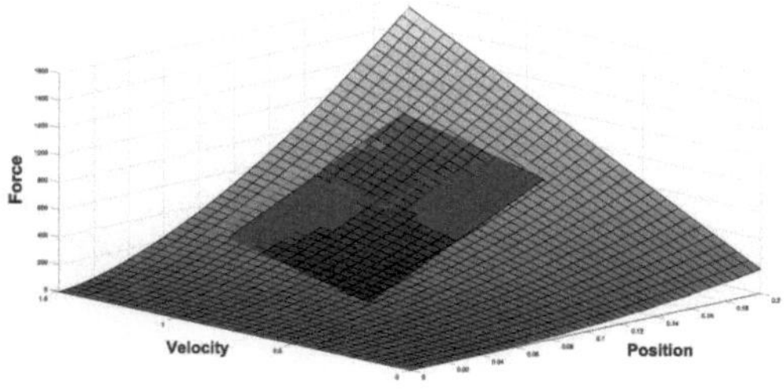

Fig. 2: Linearisation of a non-linear contact force model, describing an approximated linear behaviour around $[x_c, v_x, F_c(x_c, v_c)]^T$

Starting with equation (9) using Newton's second law $F_c = m \cdot \frac{v(t+h)-v(t)}{h}$ and an implicit integration scheme $F_c = C_1 \cdot v(t+h) + C_2 \cdot x(t+h) + C_3$ substituting $x(t+h) = x(t) + v(t+h) \cdot h$ leads to the velocity update.

$$v(t+h) = \frac{m \cdot v(t) + C_2 \cdot h \cdot x(t) + C_3 \cdot h}{m - C_1 \cdot h - C_2 \cdot h^2} \tag{10}$$

This equation should now be embedded in the LCP formulation of the dynamics simulation. Therefore, the current deflection e of penetration from the stationary state needs to be calculated.

$$e = x_{stat} - x(t) = -\left(\frac{C_3}{C_2} + x(t)\right) \tag{11}$$

Based on this difference the non-penetrating constraint can be modified, allowing to violate the constraint proportional to the constraint force λ and the current deviation e from stationary state.

$$v(t+h) + c_{cfm} \cdot \lambda = \frac{k_{er}}{h} \cdot e \tag{12}$$

Replacing the constraint force $\lambda = m \cdot \frac{v(t+h)-v(t)}{h}$ leads to

$$v(t+h) = \frac{-\frac{k_{er}}{h} \cdot \left(\frac{C_3}{C_2} + x(t)\right) + c_{cfm} \cdot \frac{m}{h} \cdot v(t)}{1 + c_{cfm} \cdot \frac{m}{h}} \tag{13}$$

Comparison of equation (10) and (13) yields the constraint force mixing parameters.

$$c_{cfm} = -\frac{1}{C_1 + C_2 \cdot h} \qquad k_{er} = \frac{C_2 \cdot h}{C_1 + C_2 \cdot h} \qquad e = -\left(\frac{C_3}{C_2} + x(t)\right) \tag{14}$$

These can now be applied to equation (6) allowing to simulate non-linear spring damper behaviour in contact normal direction given by the Jacobian J. It is worth notable, that the parameters from equation (14) are consistent with the formulation from equation (5), if $C_1 = -d$, $C_2 = -k$ and $C_3 = 0$ are chosen.

Furthermore, this constraint force mixing formulation can also be used to simulate discrete contact force models, which resolve a contact in a single time step. Typically, discrete models describe the contact behavior with help of a coefficient of restitution c_r. There are various approaches to determine this coefficient of restitution [7], but nevertheless all discrete models can be expressed in a way, that the relative velocity after the contact v_{rel}^{+} is calculated [8]. Since $k_{er} \cdot e$ is in dimension of a velocity, the relative velocity after the contact can directly be inserted in the equation system by choosing $k_{er} \cdot e = v_{rel}^{+}$ and $c_{cfm} = 0$.

The architecture of the modular framework is shown in figure 3. Because of its generic layout the contact force model can easily be exchanged, without the need of a new mathematical problem formulation. The contact force models are linearized according to the current operating point determined by the collision detection and the constraint force mixing parameters are calculated. These parameters are used to modify the LCP, see equation (6). If no linearization is required, the constraint force mixing parameters can directly be inserted to the LCP. The LCP is then solved, calculating the constraint forces and contact forces applied during the next integration step.

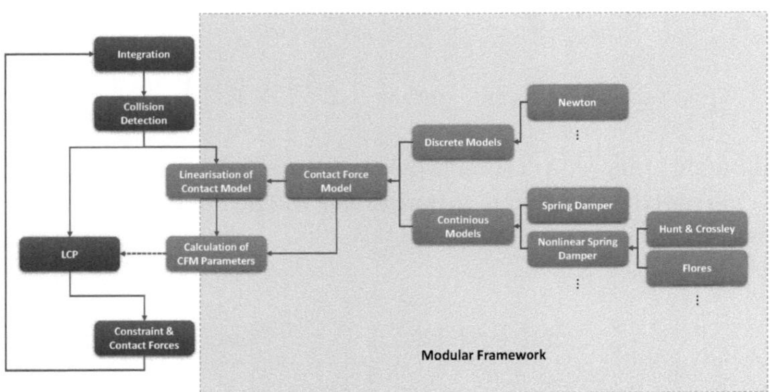

Fig. 3: Overview over the modular framework for various contact force models

5 Applications of the Framework

A falling cube is one of the simplest examples to perform for a contact simulation. The calculated contact force of a simulation with a linear spring damper simulation is shown in figure 4a and the determined contact force of a non-linear spring damper is shown in figure 4b. It can easily be seen that the spring damper model leads to vibrations of the contact force for the resting contact. A spring damper system always tries to reach its stationary state $x_{stat} = 0$. Due to the discrete integration scheme this leads to the release of some contact points and causes the LCP to oscillate.

This is one advantage of the non-linear spring damper systems. The non-linear contact force has its stationary state at $x_{stat} = -\frac{C_3}{C_2}$, leading to a small penetration, which prevents releases of contact points. Due to the adaptable definition of the hysteresis damping factor χ the non-linear spring damper model converges to the non-penetration constraint from equation (3) resulting in a stable resting contact, without any vibrations.

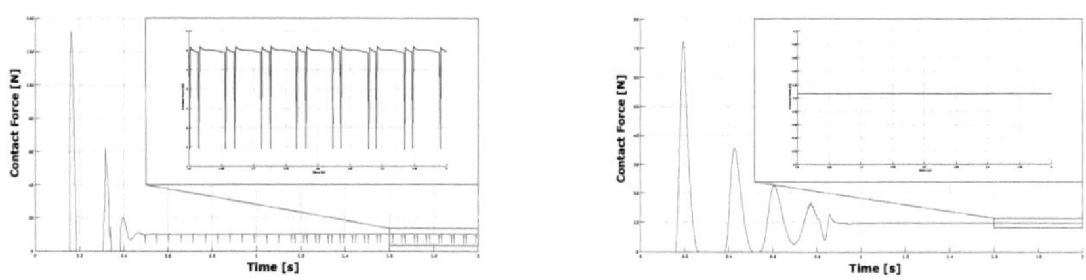

(a) Simulation with spring damper contact force model

(b) Simulation with non-linear spring damper contact force model

Fig. 4: Simulation of a falling cube

When it comes to complex contact scenarios like peg-in-hole, it is necessary to use continuous contact models. Since peg-in-hole scenarios are characterized by very small tolerances during assembly, discrete contact models will cause the peg to bounce between the limits given by the hole, see figure 5b. The same peg-in-hole assembly simulated with a spring damper contact model leads to a more smooth contact force profile, see figure 5c.

Nevertheless, a spring damper model is not the best choice for complex contact situations with many contact points. Due to discontinuities in the contact force profile (see figure 1b) and the possibility of vibrations for resting contacts, this can result in unrealistic behavior and unrealistic movements of the colliding bodies. One application for non-linear spring damper models is an interface used for building modular satellites in space [16]. In such complex contact situations more than 30 contact points can occur at the same time. Due to the continuous transition from compression to restitution phase in non-linear contact models it is now possible to simulate such complex grasping tasks with form closure.

(a) Peg-in-hole Scenario

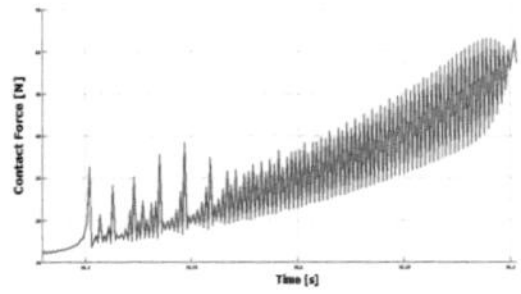

(b) Contact force profile created by simulation with a discrete contact model

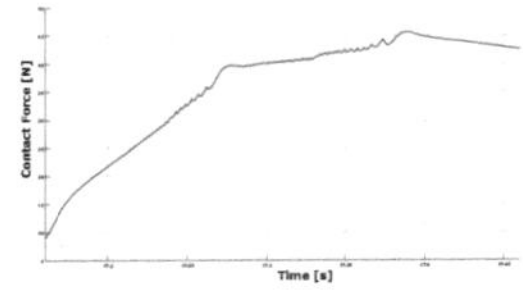

(c) Contact force profile created by simulation with spring damper model

Fig. 5: Contact forces measured during a peg-in-hole assembly

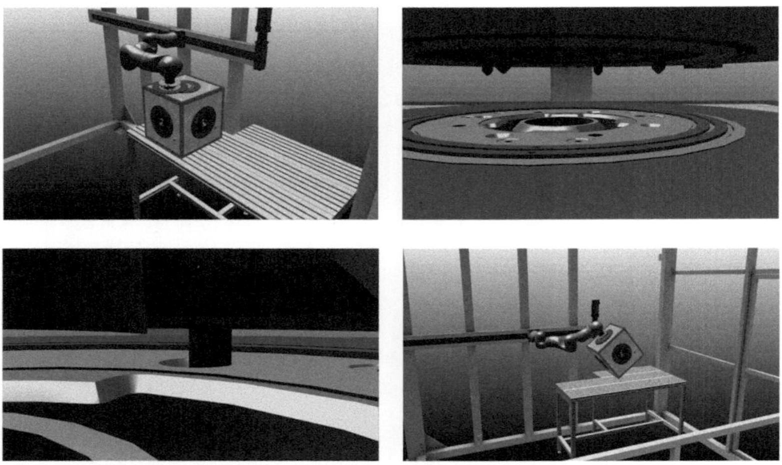

Fig. 6: Simulation of docking with interface for modular satellites [16], Mechanical Interface designed by Institute of Structural Mechanics and Lightweight Design, RWTH Aachen

6 Conclusion

The developed modular framework extends the range of validity of Virtual Testbeds, as it provides the possibility to easily integrate various contact force models in a LCP formulated dynamics simulation. This modularity is described by a generalized mathematical formulation using the constraint force mixing principle. On the one hand, this approach provides the possibility to configure the dynamics simulation to various different use cases. On the other hand, it offers an interface to easily extend the simulation by new contact force models. In future work this modularity will be extended by more relevant aspects for contact simulation, like friction models, integration schemes, as well as the underlying mathematical formulation of the dynamic simulation.

Acknowledgments

This work is part of the project "iBOSS-3", supported by the German Aerospace Center (DLR) with funds of the German Federal Ministry of Economics and Technology (BMWi), support code 50 RA 1504.

References

1. Jürgen Rossmann, Michael Schluse, "Virtual Robotic Testbeds: A foundation for e-Robotics in Space, in Industry - and in the woods," *Developments in E-Systems Engineering*, 2011.
2. Michael Schluse, Christian Schlette, Ralf Waspe, Jürgen Rossmann, "Advanced 3D Simulation Technology for eRobotics," *Sixth International Conference on Developments in eSystems Engineering*, 2013.
3. Brian Vincent Mirtich, *Impulse-based Dynamic Simulation of Rigid Body Systems.* PhD thesis, University of California at Berkeley, 1996.
4. Thomas Nicky Thulesen, *Dynamic Simulation of Manipulation & Assembly Actions.* PhD thesis, University of Southern Denmark, 2015.
5. Janete Alves, Nuno Peixinho, Miguel Tavares da Silva, Paulo Flores, Hamid M. Lankarani, "A comparative study of the viscoelastic constitutive models for frictionless contact interfaces in solids," *Mechanism and Machine Theory*, vol. 85, pp. 172–188, 2015.
6. H. Olsson, K. J. Åström, C. Canudas de Wit, M. Gäfvert, P. Lischinsky, "Friction Models and Friction Compensation," in *European Journal of Control*, vol. 4, pp. 176–195, 1998.
7. G. Gilardi, I. Sharf, "Literature survey of contact dynamics modelling," *Mechanism and machine theory*, vol. 37, no. 10, pp. 1213–1239, 2002.
8. Yu Wang, Matthew T. Mason, "Two-dimensional rigid-body collisions with friction," *Journal of Applied Mechanics*, vol. 59, no. 3, pp. 635–642, 1992.
9. Jan Bender, "Impulsbasierte Dynamiksimulation von Mehrkörpersystemen in der virtuellen Realität," 2007.
10. Thomas Josef Jung, *Methoden der Mehrkörperdynamiksimulation als Grundlage realitätsnaher Virtueller Welten.* PhD thesis, Institut für Mensch-Maschine-Interaktion, RWTH Aachen, 2011.
11. Paulo Flores, Hamid M. Lankarani, *Contact Force Models for Multibody Dynamics.* Springer, 2016.
12. Yuning Zhang, Inna Sharf, "Validation of nonlinear viscoelastic contact force models for low speed impact," *Journal of Applied Mechanics*, vol. 76, no. 5, 2009.
13. David Baraff, "Linear-time dynamics using Lagrange multipliers," in *Proceedings of the 23rd annual conference on Computer graphics and interactive techniques*, pp. 137–146, ACM, 1996.
14. Joachim Baumgarte, "Stabilization of constraints and integrals of motion in dynamical systems," *Computer methods in applied mechanics and engineering*, vol. 1, no. 1, pp. 1–16, 1972.
15. Russel Smith, *Open Dynamics Engine v0.5 User Guide*, 2004.
16. Jana Weise, Klaus Brieß, Andre Adomeit, Hans-G. Reimerdes, Michael Göller, Rüdiger Dillmann, "An Intelligent Building Blocks Concept for On-Orbit-Satellite Servcing," in *Proceedings of the International Symposium on Artificial Intelligence, Robotics and Automation in Space (iSAIRAS)*, 2012.

Development of a Methodology for the Determination of Conceptual Automated Disassembly Systems

Julius Wolff[1,a], Torge Kolditz[1], Lutz Günther[1], Annika Raatz[1,b]

[1]Institute of Assembly Technology, Leibniz Universität Hannover, Germany

[a]wolff@match.uni-hannover.de, [b]raatz@match.uni-hannover.de

Summary. At a certain point in its life cycle, a product will reach a condition where it partly or completely loses its functionality. When this happens, the disassembly has the ambition to regenerate a product-value or to enable an environmental friendly product recycling. With regard to the high workload and costs for manual labor one approach to increase the productivity of disassembly tasks is the use of automated disassembly systems (ADS). Depending on different life cycle scenarios, requirements on automated disassembly systems vary. Concerning this problem, a general methodology is developed, which enables the determination of a conceptual ADS by assigning automated modules that are processing the product disassembly. In the first place the objective of a disassembly is determined, followed by a closer investigation of the product. Thereby target components are defined, which has to disassembled. By looking at the connections between these target components suitable separation procedures are derived. Finally, modules of the automated disassembly system are determined.

Keywords: Automated Disassembly Systems, Disassembly Methodology, Life Cycle

1 Introduction

A product can reach the end of its life cycle for various reasons. It may be broken or it may be restricted in its operating life by legislations. In both cases, reaching the end of its life cycle does not mean that the product has no value anymore. Many consumer products that are not used anymore are shredded to recycle some of the materials with higher values [1]. Another possibility is, a product has not lost its functionality, so that shredding would eliminate its remaining usage potential [2]. The disassembly strategy is very different to both of these cases. In the case of component reuse a high quality disassembly process is needed, regarding a component-friendly disassembly as well as an automation concept offering promising support [3].

Furthermore, automation of a disassembly process raises its efficiency. The disassembly process cannot simply be considered as a reverse assembly process. Reasons worth mentioning are uncertainties regarding the products life cycle and different product variants [4]. Examples of these uncertainties are damages of components and connections, unknown up- and downgrades of products and different characteristics of these variants. Furthermore, the majority of products is not designed for disassembly, particularly not for automated disassembly. However, the design of automated disassembly systems (ADS) for products already existing is a challenging aspect. In many cases the design of the product is fixed and cannot be changed. Therefore, an ADS needs to provide the ability to adapt to different product variants as well as to face unplanned process disturbances caused by manufacturing uncertainties [5]. In order to design an ADS, a method for analyzing the disassembly task is presented from which the concept for an automated system can be developed.

2 Methodology

According to the claim for economical systems, the approach is to use as few different automated disassembly stations with a low number of automated modules and tooling as possible, but with the functionality to process all required disassembly tasks. To reduce the complexity of the disassembly stations, each station should provide as few different tools as possible, but enough to provide the flexibility to react to connections that cannot be detached. Furthermore, it is worthwhile, that non-destructive separation procedures are preferred over destructive separation procedures. The methodology is based on the assumption, that every disassembly station provides one automated process module (e.g. robot) for the execution of the disassembly task and eventually other handling devices that are not directly involved into the separation process. The selection of system modules is carried out in binary decision diagrams so that an automation concept can be derived from the product characteristics analytically.

2.1 Structure of the methodology

The methodology is structured in four categories (Figure 1). The first one, the determination of the objective of the disassembly task sets first boundary conditions for the processes inside the ADS. During the product analysis the components, the connections and their structural arrangement are characterized. In the following step the disassembly processes like separation and handling processes are determined. Finally, in the design of a conceptual ADS, the previous determinations are combined and the concept of an ADS can be developed.

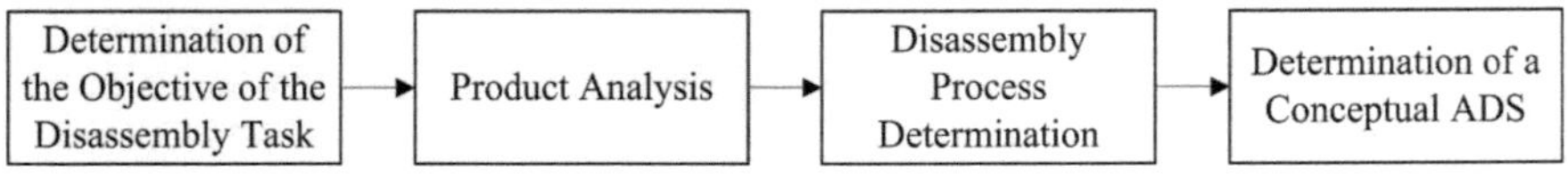

Fig. 1. Structure of the methodology

2.2 Determination of the objective of the disassembly task

Determination of the objective of the disassembly task. In a first step the permitted destruction grade and the disassembly depth have to be defined. The permitted destruction grade is mainly influenced by the life cycle scenario of a product. Table 1 considers the resulting degree of destruction and the disassembly depth depending on the life cycle scenario.

Table 1. Life cycle scenarios of the product and resulting requirements for the ADS

Life Cycle Scenario	Permitted Destruction Grade	Disassembly Depth
Maintenance	No destruction permitted (Non-destructive disassembly required)	Component separation
Repair		
Product or component reuse	Minimal destruction permitted (Semi-destructive disassembly permitted)	
Remanufacturing		
Material recycling	Destruction permitted (Destructive disassembly permitted)	Material separation

2.3 Product analysis

The product analysis is divided into four steps, as shown in Figure 2. The first step includes the investigation of the target components and their connections, from which the requirements for the ADS are derived. In the next step, component levels which are necessary for the description of the product structure are identified.

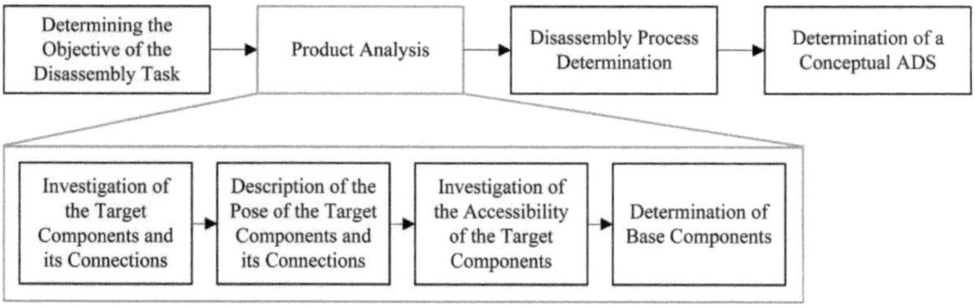

Fig 2. Product analysis for disassembly

As a first step a numbering of all target components and connections happens, so automated decision-making diagrams can be created, from which decisions can be made automatically. Connections by the same type are labelled with the same number and the numbering happens from the outer to the inner components of the product. In the following steps the number of a component is described by the counter variable k, while k_{max} describes the component with the highest number. The connection numbers are described by the counter n.

The description of the pose of the target components and their connections is essential for the separation process. To describe the pose of components to each other and related to the disassembly environment, coordinate systems (CS) are determined for every target component *k* of the product (Figure 3, left). Principally, the CS can be set at every position in the disassembly environment. However it is useful, to define a CS in the symmetry axis of a component. The z-axis of a component's CS is defined as the component's disassembly direction.

After determining the components CS, the connections between the target components have to be described. Thereby, disassembly vectors and disassembly points are defined (Figure 3, right). While the disassembly direction of a component is described by the z-axis of their CS, the disassembly vectors describe in which direction a connection has to be moved to be detached from the target components. The disassembly vectors of detachable connections can be identified by inverting their assembly directions. The vector $\vec{x}_{dis}$ defines the general disassembly direction. The magnitude of the vector describes the length of the disassembly movement.

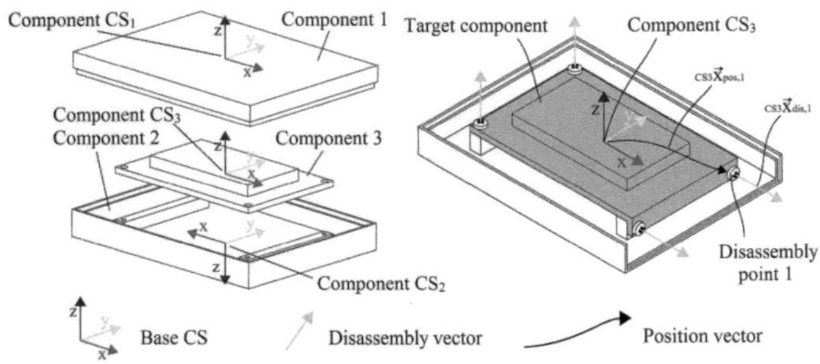

Fig. 3. Determination of component CS (left) and disassembly directions (right)

After defining the disassembly directions, the accessibility of the target components has to be investigated (Figure 4). In this paper accessibility shall be defined as the visibility and physical reachability of a component or connection. Assuming that detachable connections are accessible, if the fixed target component is accessible, the accessibility can be investigated by checking the visibility of the target component. The visibility of the component, however, is checked by considering the component from a plane aligned vertically on the disassembly vector.

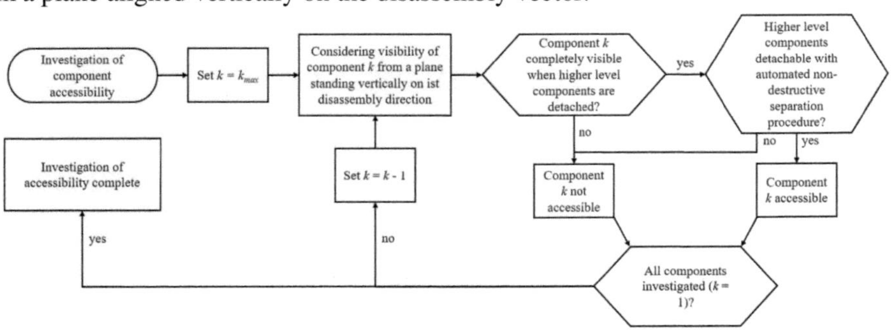

Fig. 4. Investigation of component and connection accessibility

The last step of the product analysis is the determination of base components. Base components are target components, at which other target components are mounted. If a base component can be gripped at the beginning of a disassembly task, disassembly processes can be performed, while the base component is used as the foundation for these processes. Components, identified as a base for as many separation procedures as possible are determined as base components. According to this, the only reason why there are more than one base component in products is the presence of further target components, which are not accessible in the current disassembly situation.

2.4 Disassembly Process Determination

The determination of the disassembly process is divided into two parts. First the separation processes are determined, followed by the handling processes. The description of the separation process starts by checking whether a connection is detachable. When the separation of components is performed by a destructive separation procedure (non-detachable connection), there are different possibilities to use the separation procedure. The destructive separation procedure could be used for a direct destruction of a connection or to split the material at another point of the product. The choice of where the material has to be separated is made by considering the accessibility of underlying target components. The areas where the components are separated are called separation areas. To complete a product disassembly, all target components have to be separated from each other after completing the last disassembly task. To ensure that a component of a lower component level can be disassembled, this component has to be accessible after finishing the previous disassembly task. In case that a component is not accessible, destructive separation procedures have to be performed at a higher level component. Therefore, the structure of a product has to be investigated in a first step.

Figure 5 shows an example of the determination of a separation area that ensures the accessibility of the target component. The target component is screwed onto the back cover of a case; the front cover is fixed to the back cover. The connections between the cover components are not accessible for the ADS. The disassembly direction of the target component is the direction of the z-axis; the minimal size of the separation area is shown by looking at the xy-plane of the target component.

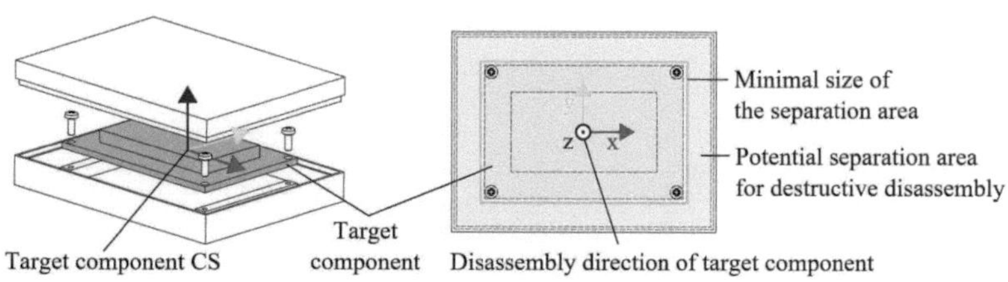

Fig. 5. Example for the determination of the separation area

In order to find suitable tooling for destructive separation processes the geometry of the separation area has to be determined. If the required separation geometry is a point (e.g. for the destruction of a screw connection), a drilling procedure can be used; if the separation area is determined by a straight line, a sawing tool can be used. Separation areas with higher complexities require a milling tool to perform the separation task. The selection processes are also reduced to similar decision diagrams as shown in Figure 4.

2.5 Determination of a conceptual ADS

The disassembly process was determined by taking the separation and handling tasks that have to be performed to complete the disassembly into account. At this stage, modules and a rough layout can be developed, based on the disassembly product and the disassembly properties. The concept of an ADS includes the number of automated disassembly stations, the motion quantities in the ADS, the required size of the workplace and automated system modules for the ADS.

The number of automated disassembly stations depends on economic and technical conditions. Examples for economic conditions are a required payback period or the maximal investment costs. Technical conditions include the technical feasible function range of single stations and requirements on the disassembly process.

The motion quantities of tools and products in and between the automated disassembly stations are determined in a way that ensures a minimal amount of handling operations. For example, it is investigated whether it is more efficient to move the product or the disassembly tools in order to perform the separation task. This depends on the structural composition of the product, the weight of the separated components and whether the required separation tool is already mounted in the concerning disassembly station.

The required size of the workspace is an important parameter of an ADS. The necessary dimensions of the workplace are determined by the disassembly and separation vectors, thus the distance that is required in each direction to disassemble components. The workspace is described by the vectors with the largest magnitude in each direction, as seen in Figure 6.

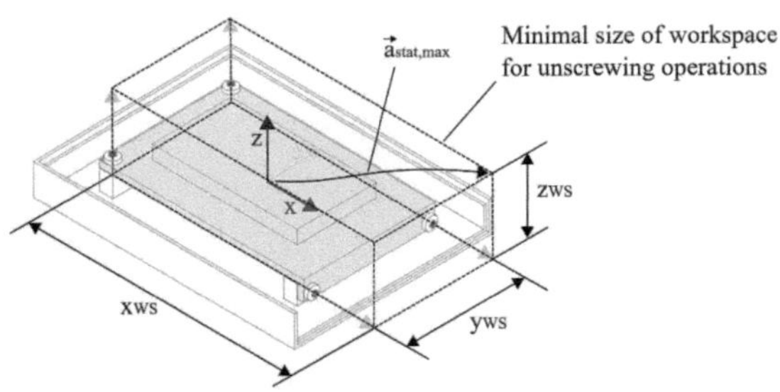

Fig. 6. Determination of the minimal size of the disassembly workspace

In a last step of this methodology, the determination of system modules takes place. First, the transfer modules are determined. Afterwards process handling modules, which perform the separation process, are determined. Transfer modules fulfil the purpose of handling units between and within the disassembly stations, but are not involved in the separation process itself. The process handling modules are selected with the aim to perform the disassembly task while needing as few degrees of freedom as possible. This approach is chosen under the assumption, that modules/robots with a higher degree of freedom cause higher investment costs.

3 Exemplary Determination of a conceptual ADS

In this section the introduced methodology shall be applied to the disassembly of LCD screens (c.f. [6]). The LCD screens, as exemplary shown in Figure 7, exist in different variants and sizes, therefore the ADS has to be able to handle different model variants. The permitted degree of destruction of the electrical components should be minimal, while the structural parts can be destructed up to a high degree.

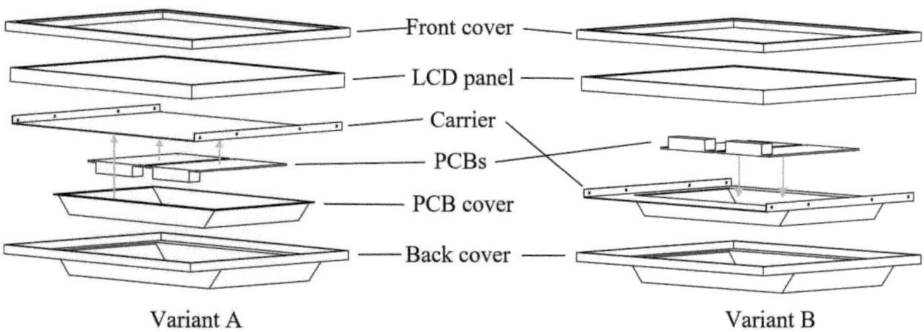

Fig. 7. Explosion view of different model variants of LCD screens (c.f. [7])

By following the steps of the methodology introduced in this paper, the system can be realized as a robot cell with one disassembly station, as seen in Figure 8. The disassembly task (Step 1) is the disassembly of LCD screens while separating different materials for recycling. In the product analysis (Step 2) coordinate systems and base components are determined. Also the description of the connections is made. In the disassembly process determination (Step 3), separation and handling processes are defined. The separation of the back cover can be performed with a sawing tool, due to the high permitted degree of destruction. The separation of the electronic parts is carried out with a screw driver though, due to the minimal degree of destruction. According to the determination of the motion quantities, the product is stationary during the performance of the separating operations, but temporary moved for reorientation. The analysis of the product characteristics and going through the decision diagrams, leads to an adapted ADS system. It contains one SCARA robot that performs the separation procedures, a second robot, equipped with a magnet gripper for the material handling, a conveyer for the transfer of LCD screens from the storage, different storage boxes, a vacuum unit for gripping and turning and a tool changing station.

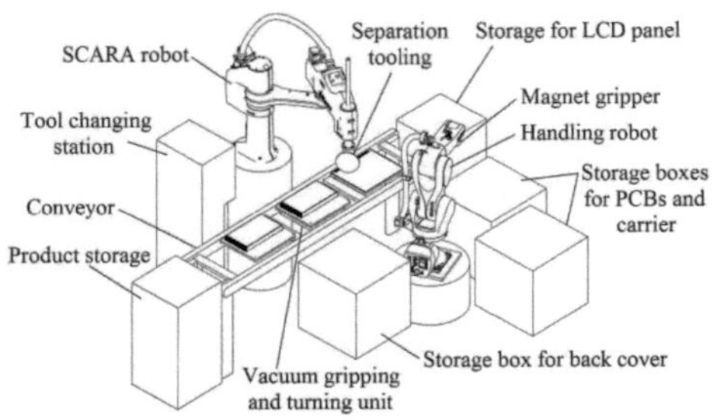

Fig. 8. Exemplary concept of an ADS for the disassembly of LCD screens

4 Conclusion and Outlook

A methodology for an automated disassembly system is presented. The method aims to generate a concept objectively from analytical decisions. Therefore, product and process characteristics are translated into simple decision-making diagrams, so that a transfer into a software tool, in analogy to existing disassembly software [7], can be realized easily. After determining the objective of a disassembly task, a product analysis follows, from which the concept of an ADS is derived. In order to estimate the separation process, target components and their connections to each other have to be defined. Finally, modules of the ADS can be determined, which is shown on an automated concept for the disassembly of LCD screens.

The results of the disassembly analysis of LCD screens as shown in chapter 3 are currently only conceptual. In further work, the methodology shall therefore be evaluated in a more realistic scenario, to verify the feasibility of the methodology.

In a future-planned software tool, the product properties are entered and a concept for a conceptual layout is generated. Once converted into a software tool, the functionality of the methodology can be expanded by adding other tools (e.g. automated tool selection [8]).

5 References

[1] Duflou, JR; Seliger, G; Kara, S; Umeda, Y; Ometto, A; Willems, B (2008) Efficiency and feasibility of product disassembly: A case-based study. CIRP Annals – Manufacturing Technology 57(2): 583-600

[2] Seliger; G (2007) Sustainability in Manufacturing – Recovery of Resources in Product and Material Cycles. Berlin, Springer Berlin Heidelberg New York

[3] Hesselbach, J., Hermann, C., Luger, T. (2007) Sustainability in Manufacturing – 3.4 Assessment of Recyclability. Berlin, Springer Berlin Heidelberg New York

[3] Gungor, A; Gupta, SM (1998) Disassembly sequence planning for products with defective parts in product recovery. Computers & Industrial Engineering 35(1-2): 161-164

[4] Seliger, G; Keil, T; Rebafka, U; Stenzel, A (2001) Flexible Disassembly Tools. Proceedings of the 2001 IEEE International Symposium on Electronics and the Environment 30-35

[5] Kim, HJ; Kernbaum, S; Seliger, G (2009) Emulation-based control of a disassembly system for LCD monitors. The International Journal of Advanced Manufacturing Technology 40: 383-392

[6] Vongbunyong, S; Chen, WH (2015) Disassembly Automation – Automated Systems with Cognitive Abilities. Cham: Springer International Publishing

[7] Ohlendorf, M., Herrmann C., Jürgen Hesselbach (2004) Simulation-based disassembly systems design, Proc. SPIE 5262, Environmentally Conscious Manufacturing III, 94

[8] O'Shea, B; Kaebernick, H; Grewal SS; Perlewitz, H; Müller, K; Seliger, G (1999) Method for automatic tool selection for disassembly planning, Assembly Automation 19(1): 47-54

Feature Recognition, Sub-Path Creation and Sequence Optimization in Systems for Processing of 2.5D Workpieces During Throughput

Denise Klose, Martin Ketels, and Thorsten Schüppstuhl

Hamburg University of Technology, Institute of Aircraft Production Technology, Denickestr. 17, 21073 Hamburg, Germany

Abstract. Product variability is a common problem for process automatization. In this work, a general approach for automatic path generation for throughput-processing of 2.5D workpieces will be discussed on the basis of two systems currently under development. Using CAD- or sensor data part-specific sub-paths for the processing machine (industrial robot) are genetared offline. They are adapted during throughput after the recognition of the specific workpiece in the robot cell. The processing order is optimized under consideration of the robot kinematic as well as the dynamic changes in environment. Therefore productivity is raised or even a feasibility enabled.

Keywords: Feature Recognition, Task Scheduling, Spray Painting, Honeycomb Filling

1 Introduction

Process automatization is driven by the need for quality improvement, reduction of work load and exposure to harmful environmental conditions as well as increased productivity. Many processing steps like welding, coating, milling etc. of 2.5D workpieces (pieces with horizontal surfaces and complex geometries such as edges and recesses or grooves, as shown in Fig. 1) can be performed by an industrial robot. Usually the robot sequentially executes a defined generated program. Manual programming is very time-consuming and therefore uneconomic for small batch sizes and high product variability. A more efficient approach is to generate the processing paths automatically based on models of the workpieces. If no CAD data are available, models based on sensor data might be generated, as in [2] and [7].

By the use of a conveyor, the machining can be performed during throughput. Often the workpieces are positioned in an arbitrarily orientation also next to each other on the conveyor. Especially long contours of large workpieces can only be treated in a short time interval, when they are located in the workspace of the robot. Thus it is disadvantageous to execute fixed, merely for one workpiece adapted programs one after another. To achieve a preferably high productivity or maybe even to enable a feasibility, a separation of the workpieces paths into sub-paths is useful, which might be executed into both directions and independent of each other. This seperation allows to switch between parts and therefore creates a new degree of freedom in the optimization of the processing sequence. In the following sections we will present two different processes which utilize this path generation and optimization.

(a) (b)

Fig. 1. Examples for 2.5D workpieces

2 Automated Spray Coating of Complex 2.5D Workpieces

This section presents a concept for automated spray coating of complex 2.5D furniture pieces in throughput currently under development in the project "Spray-One". The feature recognition and sub-path creation are described in more detail.

2.1 System Setup

The system has been introduced in [5]. It combines the flexibility of an industrial robot for coating the complex oriented geometries with the efficiency of a surface spraying machine for coating of flat surfaces. By using triangulation sensors, which are mounted above the conveyor, a model of the furniture pieces can be acquired. As shown in Fig. 2 the system is separated into pre-processing and throughput to minimize the calculation period during runtime and to enable human intervention, if there is a need to reprogramm paths or adjust parameters.

The robot paths for every known variant will be generated automatically during pre-processing. Instead of complete spray paths, separated parameterized sub-paths are generated which may be executed independently of each other and in both directions. During throughput the system will identify the workpieces, recognize their arbitrary position and orientation, and autonomously compute the optimal path over all pieces, which can also be positioned next to each other on the conveyor.

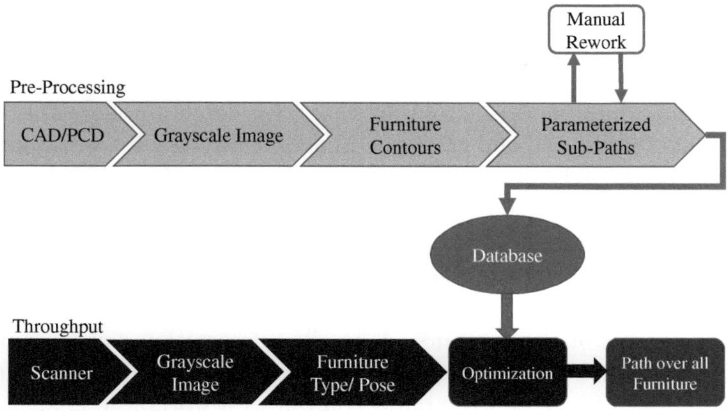

Fig. 2. System concept

2.2 Feature Recognition

The feature recognition during pre-processing is done in a grayscale range image, generated from the CAD or 3D sensor data. This simplification is feasible in work with 2.5D parts, due to the specific flat geometry no relevant data is lost. Furthermore, this format allows to solve the feature recognition with 2D image processing algorithms. In a first step the edges which border horizontal surfaces are identified by the system. They are then classified either as a single edge with an incline (Fig. 3(b)) or as part of a furrow, which has to be painted in the middle between two edges (Fig. 3(a)).

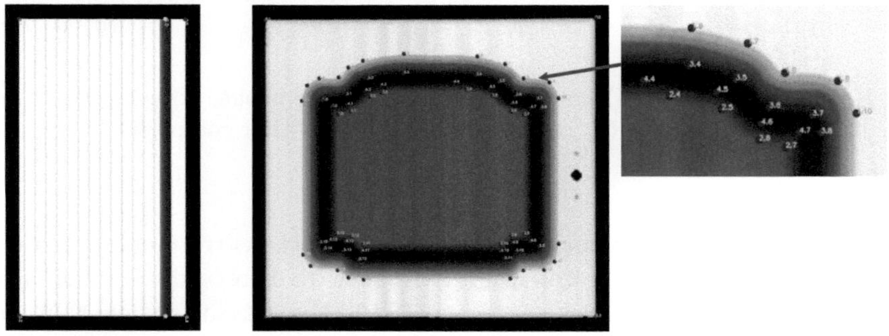

(a) (b)

Fig. 3. Greyscale images and detected contours from exemplary workpieces: (a) based on CAD data with edges (outer contour) and furrow (green points), (b) based on sensor data with edges (blue points)

To identify the edges of horizontal surfaces, a histogram based approach that uses image processing has been developed. The peaks in the histogram correspond to a horizontal surface, since the height and therefore the grayscale value of each pixel is approximately the same. Using this information and a look up table, the grayscale image is separated into multiple binary images, each representing a horizontal surface. Conventional edge detection algorithms gave no reliable results for the original grayscale images, especially for workpieces which have inclines with little slopes. In contrast, the binary images from histogram processing always give precise borders. However, two drawbacks of this method could be identified. First one beeing faulty found contours in the plane images when also white pixels appear that don't correspond to a horizontal surface because an incline is cut at the same height. The proposed solution analyzes the neighboring regions of the contours. A detailed description can be found in [5]. The second problem is the faulty detection of duplicates, which mostly accurs at the outer contour. Therefore, the outer contour of a workpiece is identified and treated separately prior the plane image creation.

(a) (b) (c) (d)

Fig. 4. Contour Finding: (a) grayscale image, (b) plane image, (c) comparison image with dilated outer contour, (d) contours in plane image after comparison

Figure 4(a) shows the generated grayscale, 4(b) the corresponding plane image of a furniture piece generated from CAD data. To disregard the small furrows in the horizontal surface also a spectrum around the grayscale value that corresponds to the horizontal surface is mapped to white in the binary image. In the plane image the contours of detected closed contours are shown as blue points. Each of them borders a white rectangle, thus consists of four segments. As it is obvious the shorter segments overlap with the outer contour of the furniture and only the long segments, which correspond to a furrow are of interest. To solve this problem, a comparison image has been introduced. Into this image the binary plane images are copied after they have been processed. The first one is the dilated outer contour (Fig. 4(c)). The contour segments end- and middle-pixels, which were found in the plane image, are matched with the corresponding ones in the comparison image. If they are white, the segment is deleted. Thus the segments that are part of a furrow remain (Fig. 4(d)). As it is obvious, these contours are open. Their start- and end-points are not the same.

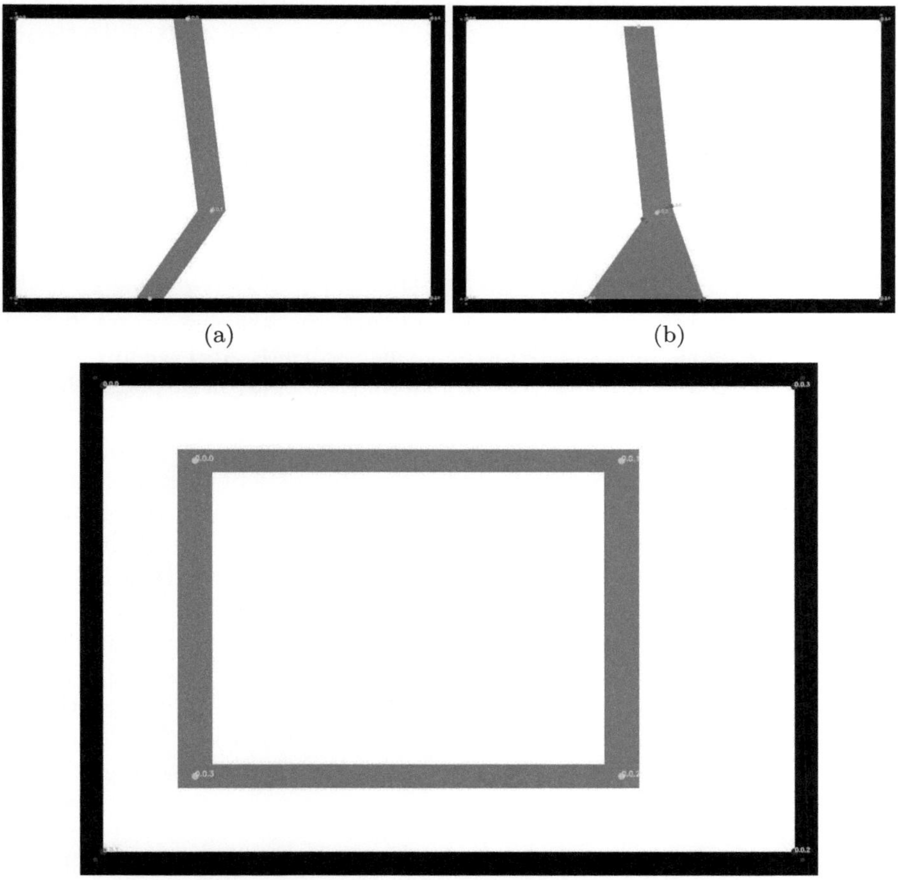

(a) (b) (c)

Fig. 5. Sub-Path generation: (a) furrow (green) with two segments and outer contour splitted;(b) furrow (green), edges of recess (red) and outer contour splitted; (c) furrow (green) with four segments and outer contour not splitted (blue)

When having evaluated the closed and open contours which border horizontal surfaces, it is determined whether they are part of a recess or of a furrow. Contours segments which are part of a recess have an incline which has to be painted with an oriented tool, whereas a furrow has to be painted in the middle between two segments and the tool is orientated vertically. The evaluation, if a pair or multiple pairs of segments build a furrow, is done in each plane image, but before the contours have to be prepared. The segments of the contours are sorted in a defined order and parameters like the side of the surface and segment length are determined. To recognize a furrow, the segments are checked pairwise. In an open contour itself and also between contours of a plane image. To build a furrow the segments must have approximately the same length, be approximately parallel and the connecting line between the middle points of the segments mustn't have white pixels, which means there is no surface between the segments. For achieving also connected furrow-segments as it is shown in Fig. 5(a) and 5(c) the order of the contour-segments and the sequence of creating the furrows is relevant. Figure 5(b) shows a recess. A part of it is recognized as furrow, whereas the remaining contour segments are part of an edge of the recess.

2.3 Sub-Path Creation and Sequence Optimization

For the sequence optimization during throughput, the contours are splitted into sub-paths depending on the angle between the segments. Also the angle for the tool and an offset is calculated. The displacement for the offset points depends on whether the point is an end point of the sub-path or an intermediate point. End points are shifted perpendicular to the segment whereas intermediate points are shifted on the bisector of the angles between the segments. The outer contour in Figures 5(a) and 5(b) was separated, thus the offset points (the smaller red dots) are shifted perpendicular. By setting another threshold angle in turn the outer contour hasn't been separated (Fig. 5(c)). In this case the offset is shifted on the bisector (smaller blue points).
During throughput the type and position of the workpieces on the conveyor are estimated based on the range images generated from scan data (Fig. 6). After loading the corresponding sub-paths (effective tasks) from the database the system should adapt them according to their pose and sort them into an optimal order to autonomously compute the optimal path over all pieces on the conveyor. The optimal order also depends on the paths which connect the effective tasks (intertasks).
In static cases solutions for problems of combinatorial optimization, like the "Traveling Salesman Problem" exist [4]. A traveling salesman should visit all given cities (effective task), with known distances (intertask) on the shortest way. As described in [1] also modified problem formulations exist. In the regarded case the TCP of the robot doesn't only have to visit one point. The effective task consists of a painting movement, like a linear movement. Thus the starting- end end-points are not the same and might be chosen. Consequently the quantity of possible solutions expands. Another constraint comes with the robots kinematic. Its arm length and joint angles are finite and therefore not all movements, which are calculated in task space are feasible. The problem has to be regarded in joint space. Often multiple solutions exist to reach a point. Due to this aspect, the number of possible solutions for the optimization problem rises a second time. For static robot applications solutions for scheduling of robot tasks are already implemented, which involve constraints like the robots

kinematic, multiple solutions or painting quality [1, 6]. However in the present circumstances it is important to consider that the objects are in movement and only positioned for a finite time in the workspace of the robot, thus every subpath might only be threated in a defined time interval. Caused by the kinemtic and depending on the chosen start time for an effective path, different movements have to be fulfilled. Also the intertask vary, thus the regarded problem changes continious with the time. The number of possible solutions rises. Therefore actualy an approach to solve such an dynamic traveling salesman problem is in development, which includes the robots kinematic and the temporary change of the problem.

Fig. 6. Grayscale image of furnitures lying on the belt.

3 Automated Honeycomb Potting

Another application which involves processing of 2.5D workpieces is the manufacturing of honeycomb sandwiches. A system concept for automated honeycomb potting is presented in this section.

3.1 Backround

Sandwich structures with honeycomb core are widely used in aeronautic industry, mostly due to their excellent weight-to-rigidity ratio. Possible applications are numerous [3], the most common usage being as interior paneling. For transferring force, to locally increase the stiffness and for edge sealing some regions of the core material are potted with epoxy resin. As common in airplane manufacturing the sandwich panel production consists mainly of manual processes. Automatization yields not only high potentials in view of productivity but also in respect to achievable product quality. Potting material is currently pressed either by hand or with a dispensing gun into the core, avoiding entrapped air and partially filled honeycombs requires a lot of experience. Known efforts for automatization are limited to robot guided and controlled dispensing gun.

3.2 System Setup

A new concept for automated honeycomb potting is currently in development. Major problems are high product variability and frequency of changes, which are general characteristics in the aeronautic industry. The path generation will reduce programming effort and make an automated processing feasible. The process itself is very similar to the spray painting, with a tool that is moved in continuous manner and perpendicular to the surface structure of the 2.5D-workpiece. Therefore, many results from [5] can be incorporated in this concept. The pre-processing will use the CAD-data of the parts. These contain the final dimensions and the regions to be filled, either as solids (for 2D parts) or as templates (for 2.5D parts). The nozzle has a fixed diameter, and wider regions are separated in multiple sub-paths. These can be processed independently and from any approach direction. Overlaps are also possible, since the width of a region is not necessarily a multiple of the nozzle diameter. These sub-paths are generated offline and stored in a database. The possibility to place the workpieces in any order and pose on the conveyor simplifies the integration of the automated potting into the otherwise manual process chain. The recognition of the parts is less of an issue as in the spray coating, since the parts are manufactured in tools (for later hardening in an autoclave or heated press), on which tags or markers can be mounted. The markers can also be used for pose estimation in the recognition phase. This rather rough knowledge of the macro-structure will be sufficient for the determination of the optimal order of sub-paths.
The main advancement to the solution for the spray coating will be in a new module for online path adaption. The micro-structure (position and shape of single honeycombs) can effectively only be measured during processing. The nozzle needs a certain overlap to ensure complete filling of a honeycomb and the robot path needs to adjust accordingly. This principle is illustrated in Fig. 7. A sensor which is mounted ahead of the nozzle scans the core layer, the adaption algorithm then adds perpendicular motion to the current sub-path if necessary. In corner cases this could require the generation of new sub-paths, if single honeycombs would be left out after adaption. Therefore a feedback loop will be implemented, which can add sub-paths to the processing pipeline.

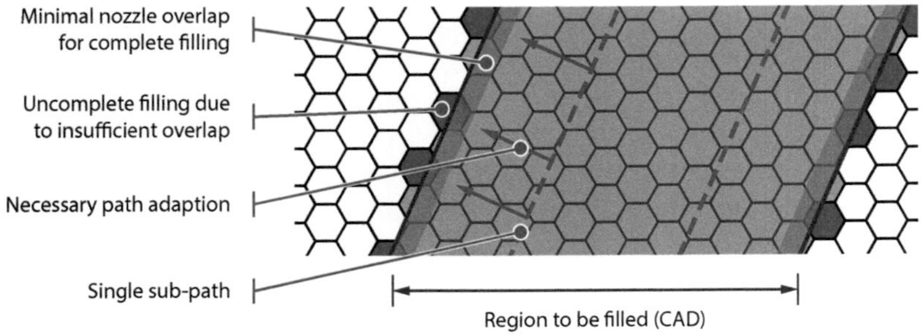

Fig. 7. Path adaption for complete potting of honeycombs

4 Conclusion

This paper treats systems that process 2.5D workpieces of small batch sizes during throughput by machines like industrial robots. The main problems, like path generation based on geometric models, path adaption and sequence optimization in a temporally changing environment are elaborated. As exemplification two applications were presented, that are actually under development.
With the presentation of the concept for automated spray coating a developed method for feature recognition and sub-path creation, that makes use of the characteristic flat geometry of 2.5D workpieces, was introduced. Additionally the challenges of the sequence optimization in robotic systems with dynamic environments, which has to be solved in future work, were pointed out. Due to the robot's kinematic and especially the movement of the workpieces the number of possible solutions for an overall path rises significantly compared to static problems. The concept for automated honeycomb potting demonstrates a second exemplary application for the sequence optimization during throughput with an additional online path adaption.

Acknowledgements

This work is supported by the German Federal Ministry for Economic Affairs and Energy (BMWi) with the projects "SprayOne" and "RoboFill".

References

1. Alatartsev, S., Stellmacher, S., Ortmeier, F.: Robotic task sequencing problem: A survey. Journal of Intelligent & Robotic Systems **80**(2), 279–298 (2015)
2. Gasparetto, A., Vidoni, R., Pillan, D., Saccavini, E.: Automatic path and trajectory planning for robotic spray painting. In: Robotics; Proceedings of ROBOTIK 2012; 7th German Conference on, pp. 1–6. VDE (2012)
3. Herrmann, A.S., Zahlen, P.C., Zuardy, I.: Sandwich Structures Technology in Commercial Aviation, pp. 13–26. Springer Netherlands, Dordrecht (2005). DOI 10.1007/1-4020-3848-8_2
4. Jungnickel, D.: Graphs, networks and algorithms, vol. 5. Springer-Verlag Berlin Heidelberg (2005)
5. Klose, D., Schueppstuhl, T., Osthues, D.: Automated spray coating of complex 2.5 d workpieces in small batch sizes with complex geometries in throughput by an innovative machine combination. In: ISR 2016: 47st International Symposium on Robotics; Proceedings of, pp. 1–6. VDE VERLAG GmbH (2016)
6. Kolakowska, E., Smith, S.F., Kristiansen, M.: Constraint optimization model of a scheduling problem for a robotic arm in automatic systems. Robotics and Autonomous Systems **62**(2), 267–280 (2014)
7. Pichler, A., Vincze, M., Andersen, H., Madsen, O., Hausler, K.: A method for automatic spray painting of unknown parts. In: Robotics and Automation, 2002. Proceedings. ICRA'02. IEEE International Conference on, vol. 1, pp. 444–449. IEEE (2002)

Design and Evaluation of a Multi-Agent Software Architecture for Risk-Minimized Path Planning in Human-Robot Workcells

Tobias Werner, Dominik Riedelbauch, and Dominik Henrich

Chair for Robotics and Embedded Systems,
Universität Bayreuth, D-95440 Bayreuth, Germany,
tobias.werner@uni-bayreuth.de,
http://robotics.uni-bayreuth.de

Abstract. Close quarters human-robot collaboration promises the integration of human flexibility with robot precision, endurance, and strength. However, such collaboration requires a monitoring solution for the shared human-robot workspace to drive the robot manipulator. Recent research has already developed a variety of individual software components for monitoring solutions, from object detection over reconstruction to path planning. In our contribution, we implement existing components for multi-camera reconstruction, risk estimation and path planning in the form of concurrent agents and we connect these agents through a shared blackboard in order to realize risk-minimal path planning for a robot manipulator. We discuss our software architecture, and we evaluate the final software in an example application of human-robot collaboration. In conclusion, our contribution enables risk-minimal path planning on an industrial manipulator at a 10 Hz update rate.

Keywords: risk-minimized path planning, blackboard path planner, software architecture, shared human-robot workcells

1 Introduction

For past decades, robot manipulators have proven their worth in industrial automation due to their strength, endurance, and precision. Humans, in contrast, excel in cognition and can quickly adapt to unexpected or unplanned situations. Therefore, it seems worthwhile to combine the individual advantages of humans and robots in a shared workspace. Prospects include increased process flexibility and reduced maintenance downtimes.

Close quarters collaboration between humans and robots is only acceptable if robots can perceive and evade human coworkers in the shared workspace. Our particular goal is to find a real-time, risk-minimized path for a robot manipulator in a shared workspace under the presence of moving humans or other a priori unknown obstacles. See Figure 1 for an example application.

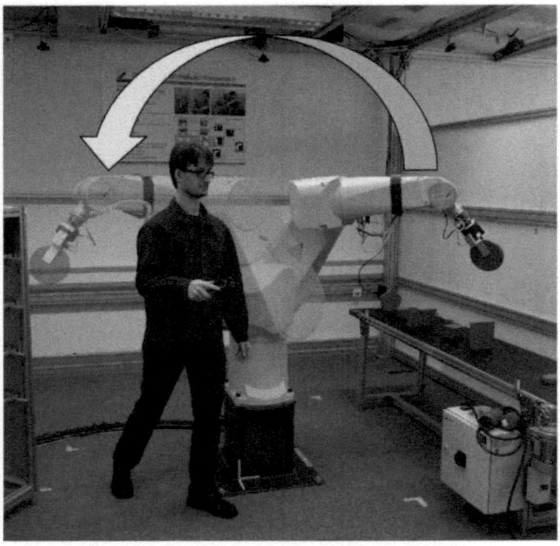

Fig. 1. Our example application: A Stäubli RX130 robot manipulator is to transfer items from the right-hand conveyor belt to the left-hand tool shelf. To minimize the risk of a collision with humans or a priori unknown obstacles in the shared human-robot workspace, the robot perpetually has to adapt its trajectory in real-time.

More formally, we must find a new path $p_t : [0,1] \rightarrow \mathbb{R}^n$ in the n-dimensional configuration space from the current robot position $p_t(0) = q_{t,s}$ to a set goal $p_t(1) = q_e$ at regular update steps $t \in \mathbb{R}$. We express risk for the current update step through a risk function, $r_t : \mathbb{R}^n \rightarrow \mathbb{R}$. This function considers detrimental factors, including the probability of object presence within the robot geometry for given joint coordinates. In theory, we seek a path that globally minimizes the risk, $p_t = \mathrm{argmin}_p(\int r_t(p(\lambda))\,\mathrm{d}\lambda)$. In practice, it is impossible to find a globally risk-minimal path. We instead strive for a reasonably risk-minimized path.

In the following, we contribute an alternative software architecture that is tailored specifically to the demands of risk-minimized path planning. Our software architecture is both light-weight and efficient: On the one hand, we rely on concurrent (i.e. distributed or threaded) computations where possible. On the other hand, we combine a multi-agent paradigm with communication through a blackboard to minimize coupling and to maximize cohesion of individual software components. This in turn allows for a light-weight interfacing of software components even over thread or network boundaries. Finally, we choose existing software components that are well-suited for risk-minimized path planning.

The remainder of our contribution is structured as such: Section 2 presents related work, both in terms of software architectures and in terms of software components. Section 3 gives a high-level overview over our software architecture for risk-minimized path planning. Section 4 highlights specific software components within our software architecture. Section 5 evaluates our architecture in an example use case from human-robot collaboration. Section 6 reviews our contribution and concludes with an outlook on future work.

2 Related Work

In general, research proposes manifold reactional and behavioral software components for workspace monitoring in the presence of a priori unknown objects (e.g. speed control and path planning [9]). Each software component requires a specific type of environment representation (e.g. point clouds [31], voxel spaces [25]). The environment representations in turn originate from individual sensors (e.g. artificial skins [26], depth cameras [17], color cameras [11]), or from fusion of data over multiple sensors (e.g. multi-camera systems [19] that find visual hulls [15] or photo hulls [4]). Throughout all software components, there are two distinct variants: approaches either follow a Boolean (e.g. [13]) or a probabilistic (e.g. [23]) paradigm. Boolean variants use discrete input (e.g. object silhouettes) to derive Boolean environments (e.g. object volumes) for collision-free path planning. Probabilistic variants use continuous input (e.g. probabilities for object presence) to derive probabilistic output (e.g. 3D certainty grids) for optimized path planning. Notably, probabilistic variants can incorporate sensor errors, system latencies, or trajectory risks (e.g. [12], [3]).

Each of the above software components runs an intricate and potentially expensive algorithm. To control a robot manipulator, a real-world monitoring solution has to interface a variety of such components while satisfying soft real-time constraints. Existing solutions for this interfacing split into three distinct categories: robot system architectures, knowledge data bases, and world models. Robot system architectures (e.g. ROS [21]) focus on flexible data exchange over software components, but usually prefer stability and extensibility over low-level performance. Knowledge data bases (e.g. RoboBrain [24]) expose extensive semantic knowledge through relational queries, yet do not care much about real-time system latencies. Finally, world models (e.g. OctoMap [31]) enable efficient access to low-level geometric data, but do not explicitly consider integration with threaded or distributed software components.

At implementation level, software architectures tend to follow established software design paradigms (see [8]). As risk-minimized path planning involves software components with expensive and intricate algorithms, design paradigms for threaded or distributed workload are of particular relevance. To achieve implicitly safe, efficient, and extensible parallelization, research proposes a paradigm of minimized coupling and maximized cohesion (see [14]). Multi-agent systems (see [27]), often combined with blackboard communication schemes (see [10]), are a common implementation of this paradigm. In robotics, distributed blackboards predominantly synchronize shared knowledge over mobile robot agents (e.g. [6]) and manage light-weight software components within a single mobile robot (e.g. [28]). Opposed to approaches for mobile robots, ENACT (see [30]) is a software architecture designed specifically for robot manipulators. ENACT successfully combines multi-agent software components with a blackboard in form of a world context to solve a variety of sub-symbolic and symbolic tasks, including collision free path planning or physically grounded pick-and-place operations. Therefore, applying the same paradigms to the computationally expensive task of risk-minimized path planning seems promising.

3 Software Architecture

Our contribution is a novel software architecture specifically tailored to risk-minimized path planning. As such, our contribution stands in contrast to above, more generic architectures. The remainder of this section presents a summary of our overall software architecture, while the next section elaborates on individual software components within our architecture.

At the top-most layer, our software architecture represents a generic pipeline for path planning: A source stage acquires sensor data through a multi-camera system (see Figure 2), a subsequent processing stage derives path segments from sensor data, and a terminal sink stage dispatches path segments to the robot manipulator for execution. Our software architecture implements all three stages by means of a distributed processing model. The source stage distributes over a series of *Camera PCs*, each occupied with expensive preprocessing of images from a single camera. The processing stage concurrently performs risk-minimized path planning over preprocessed sensor input on a single *Server PC*. Finally, an independent *Controller PC* drives a robot manipulator with resulting path segments. This distribution model follows the paradigm of minimized coupling and maximized cohesion. Calculations on individual PCs are mostly independent and thus do not couple, while the tightly coherent state of path planning remains local to the Server PC. Transfers in-between PCs in turn require but light-weight and efficient interfaces. In the context of our overall software architecture, distribution yields two major advantages: On the one hand, distribution enables us to offload expensive preprocessing to separate PCs, a vital contribution to real-time update rates. On the other hand, distribution hardens our architecture against failure of individual software or hardware components.

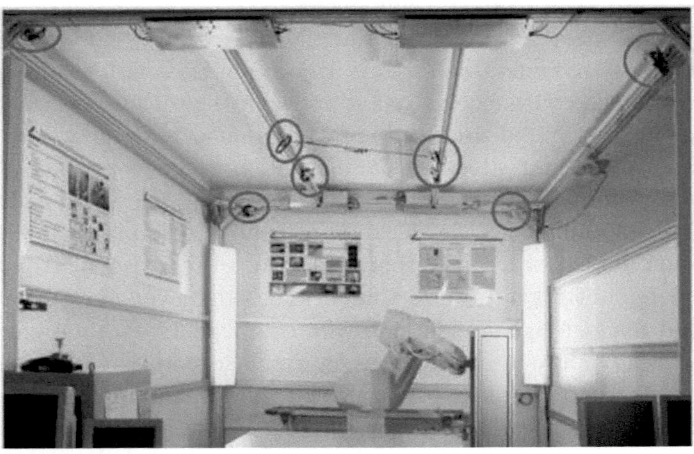

Fig. 2. To detect humans or obstacles within the shared human-robot workspace, we use a multi-camera system. This system consists of multiple intrinsically and extrinsically calibrated color cameras (red circles).

At a layer below the distributed model, our software architecture employs threaded software components to take advantage of the multi-core CPU in the Server PC. A standalone *3D Reconstruction* component performs sensor fusion on any camera images that Camera PCs transmit over network. The other software components on the Server PC are concerned with path planning and path execution. These software components collaborate under a multi-agent paradigm: A blackboard holds path suggestions, which both path planning and path execution agents read and write to. Path planning agents evaluate current results of sensor fusion to derive new path suggestions, to modify existing suggestions, or to discard stale suggestions. While the choice of planning agents — and of respective software components — is flexible, there are three mandatory agents: Threaded *Path Annotation* synchronizes risk estimates of stale blackboard suggestions to current fusion results. *Path Selection* uses annotated risks to perpetually choose a risk-minimized path from robot position $q_{t,s}$ to joint-space goal q_e. Finally, *Path Execution* pulls pending segments of the currently selected path from the blackboard and transmits these to the *Controller PC* for execution by the robot manipulator. Figure 3 illustrates software components in our software architecture and shows data flow between individual components.

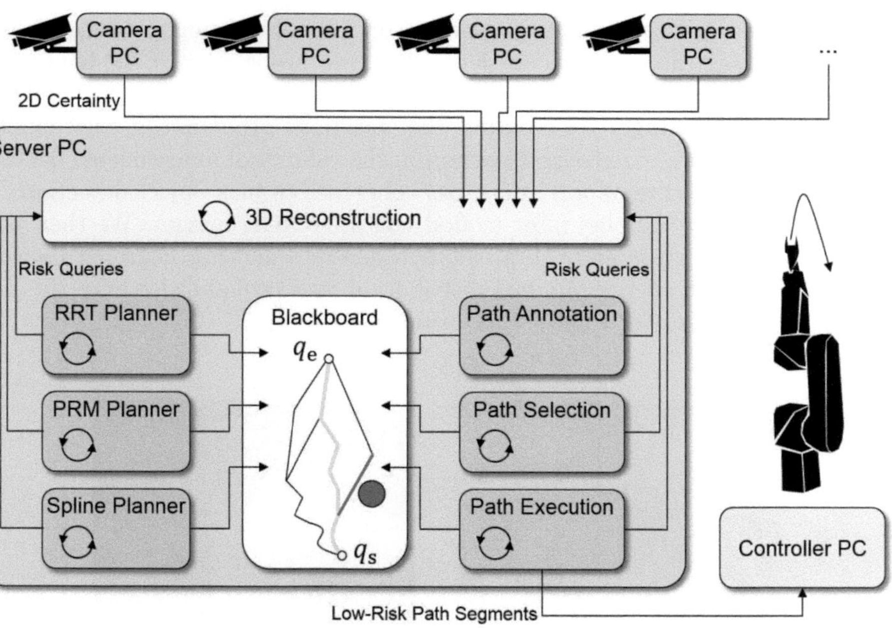

Fig. 3. Overview over software components and data flow in our software architecture. *Camera PC*s, the *Server PC*, and the robot *Controller PC* follow a distributed model, while arrow cycles indicate multi-threaded software components on the Server PC. The shared blackboard currently holds three path suggestions. The red path segment implies high risk due to proximity of an exemplary object. The green path yields minimized risk and thus has been chosen for execution.

4 Software Components

In this section, we provide details on some of the software components that have been introduced in the preceding section. Due to limited scope, we refer to original sources where appropriate. As we desire to find a risk-minimized robot path, we are restricted to software components that can handle probabilistic data. In particular, probabilistic data permeates our entire pipeline from image preprocessing on the Camera PCs to path selection on the Server PC. We perform a deterministic choice but at the last possible step, directly before dispatching a single path segment to the Controller PC for subsequent execution.

The first software component in our probabilistic pipeline performs image preprocessing on distributed Camera PCs and transforms RGB images from the multi-camera system into 2D certainty maps. The respective algorithm consists of two sequential stages: In the first stage of preprocessing, a neural network estimates foreground object probabilities through background subtraction as proposed in [7]. Over time, the neural network collects per-pixel foreground-background information and models pixel statistics within its edge weights. The neural network then compares incoming input images with the current background model to derive independent per-pixel foreground probabilities. To avoid forgetting static objects (i.e. to suppress the sleeping person problem), we extend the original implementation by a modified conditional learning method, where a variable learning rate depends on the background probability of the final pixel. In the second stage of preprocessing, we exploit a priori knowledge to enhance results of the first stage. We conclude that per-pixel probabilities must exhibit spatial and temporal homogeneity except in the vicinity of image edges. We thus consider spatial and temporal inter-pixel coherence to bias object detection. To this end, we apply a Sobel filter to find edges on RGB images. We then pass edges from the current and the previous frame into another neuronal network to approximate an energy-minimal distribution over probabilities from the first preprocessing stage. See Figure 4 for example results.

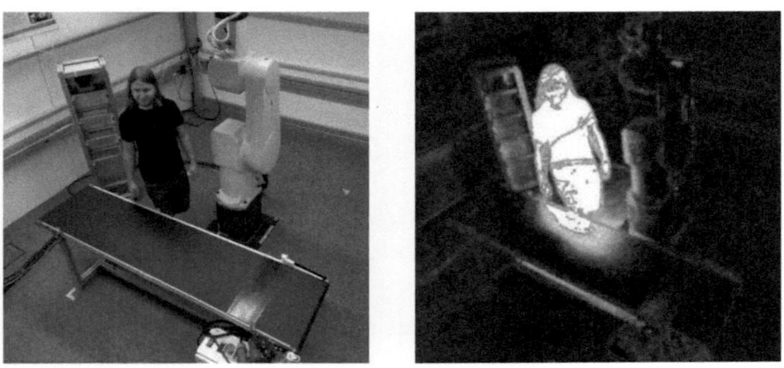

Fig. 4. Left: High-resolution image from the multi-camera system. Right: 2D certainty map with foreground probabilities. Note that these images do not show additional a priori knowledge, including the current manipulator volume or static workspace objects.

The Camera PCs transfer final 2D certainty maps to the Server PC over TCP/IP network. Quantification and run-length encoding avoid network congestion, while an explicit pull paradigm prohibits unnecessary transfers.

Concurrently to pending network transfers, the software component for 3D reconstruction performs sensor fusion on a synchronized set of 2D certainty maps. The goal of sensor fusion is to build a 3D certainty map with probabilities of object presence for all positions within the workspace. Our implementation closely follows the hierarchical and incremental paradigms presented in [29]: Build quadtrees over input maps, and generate an octree over the workspace through efficient quadtree lookups. During this process, track changes in input maps to maintain unchanged octree branches for efficiency reasons. The original fusion approach [29] only considers binary object presence, so we must adapt this strategy to integrate probabilistic data. To this end, we maintain upper and lower bounds on leaf probabilities within inner quadtree nodes. For each node of the octree in the workspace, we consider camera projections into quadtrees to determine upper and lower bounds of 2D probabilities. At this point, we assume independent probabilities over 2D certainty maps, which allows us to find 3D probabilities by multiplying all 2D probabilities. Finally, we split octree nodes once the difference between lower and upper probability bounds exceeds a set threshold. The resulting octree encodes 3D object probabilities over the workspace. Subsequent software components do not access this octree directly. The 3D reconstruction component instead exposes a risk query that samples the octree over the robot volume which corresponds to a given configuration-space position. This risk query acts as an integral part of the risk function r_t.

The final software components concurrently handle path planning and path execution while collaborating over the shared blackboard. We initialize our blackboard with a single, directed path segment from the robot position $q_{0,s}$ to the goal position q_g. Path planners then iteratively add, execute, or delete segments until the robot has reached its goal position. We have selected and implemented three run-of-the-mill path planners for use in our software architecture: The *RRT Planner* uses rapidly-exploring random trees and adheres to the original RRT proposal [16]. The *PRM Planner* searches for paths on a probabilistic roadmap and closely follows the implementation in [9]. The *Spline Planner* generates paths by applying bezier splines of varying parameters. All three path planners have been adapted to collaborate through the blackboard mechanism. Notably, planners concurrently pick segments from the blackboard and apply their respective planning strategy to determine a more risk-minimal alternative — as estimated through per-node risk queries on 3D reconstruction. Virtues of individual planners govern segment picking. For instance, the RRT planner is prone to picking long segments, while the spline planner concentrates on sharply bent segment pairs. In the end, collaboration of path planners results in a series of path suggestions on the blackboard. *Path Selection* implements a common Dijkstra variant for directed acyclic graphs to perpetually choose a risk-minimal path from these suggestions (e.g. as in [5]). The remaining, mandatory software components realize intuitive functionality and are not discussed here.

5 Experiments

We have evaluated our software architecture in the form of a prototype implementation. In our prototype, a multi-camera system monitors the workspace around a Stäubli RX130 robot with eight inexpensive, consumer-grade Logitech C930e cameras. We distribute preprocessing of incoming 1920x1080 images to 2D certainty images over one PC per camera, where low-end NVIDIA GTX750 GPUs carry most of the workload. A Server PC with an Intel Core i5-4670 CPU polls preprocessing results over Gigabit Ethernet and performs both reconstruction and path planning. 3D reconstruction runs in parallel over two threads, while another six threads correspond to path planning agents. With this setup, we are able to feed risk-minimized path segments into the robot at a 10 Hz rate.

In the context of system performance, we deliberately use a 10 Hz rate to delimit soft real-time capabilities. Three factors contribute to this choice: First, we estimate that overall system latency (e.g. due to USB stacks on the camera PCs, network latency, or processing in the — albeit dated — RX130 CS7 robot controller) has the same order of magnitude. An improvement over the indicated 10 Hz software rate will thus not peculiarly affect overall performance. Second, a 10 Hz update rate is well below the human reactions rate of at most 5 Hz, which enables seemingly instant reactions to human actions within the workcell. Third and last, typical human actions within our workcell imply a limb displacement of at most 2 m/s (see [22] for walking and [18] for pointing gestures). Artificially biasing risk within a respective 20 cm boundary around the current robot volume takes such movements into account without impairing large-scale path planning. Still, one must note that our software architecture is a monitoring solution for risk-minimized path planning and not a safety solution. Additional measures (e.g. an artificial skin [2], a soft manipulator [1], or explicit human tracking [20]) are required to ensure the safety of humans within the shared workcell.

6 Conclusion

In the preceding, we have contributed a software architecture that enables risk-minimized path planning on a robot manipulator. Path planning executes at a soft real-time rate of 10 Hz. Our software architecture achieves this through distributed preprocessing of 2D certainty maps, and through a blackboard paradigm for multi-agent reconstruction, path planning, and robot control.

In future work, we intend to integrate additional safety measures — including an artificial robot skin — into risk minimization. We also plan to integrate risk-minimal path planning into a more involved scenario of flexible human-robot collaboration: In this scenario, task planning can incorporate path risks, which enables the robot choose a risk-minimal goal from a selection of pending tasks.

References

1. A. Albu-Schaffer et al.: *Soft robotics*, Robotics and Automation Magazine, vol. 15, no. 3, 2008.
2. F. Bergner, E. Dean-Leon, G. Cheng: *Event-based signaling for large-scale artificial robotic skin - realization and performance evaluation*, Intelligent Robots and Systems, 2016.
3. L. Blackmore et al.: *A probabilistic approach to optimal robust path planning with obstacles*, American Control Conference, 2006.
4. A. Broadhurst, R. Cipolla: *A Statistical Consistency Check for the Space Carving Algorithm*, British Machine Vision Conference, 2000.
5. T. Cormen et al.: *Introduction to Algorithms, Section 24.3: Dijkstra's algorithm*, ISBN 0-262-03293-7, MIT Press, Second Edition, pp. 595-601, 2001.
6. G. Brzykcy et al.: *Multi-agent blackboard architecture for a mobile robot*, Intelligent Robots and Systems, 2001.
7. D. Culibrk et al.: *Neural Network Approach to Background Modeling for Video Object Segmentation*, Transactions on Neural Networks, 2011.
8. E. Gamma et. al.: *Design Patterns. Elements of Reusable Object-Oriented Software.*, ISBN 978-0201633610, Prentice Hall, 1994.
9. T. Gecks: *Sensorbasierte, echtzeitfähige Online-Bahnplanung für die Mensch-Roboter-Koexistenz*, PhD thesis, Universität Bayreuth, 2011.
10. B. Hayes-Roth: *A blackboard architecture for control*, Artificial Intelligence, vol. 26, no. 3, pp. 251-321, July 1985.
11. S. Kuhn: *Multi-view reconstruction in-between known environments*, Technical report, Universität Bayreuth, 2010.
12. B. Lacevic, P. Rocco: *Towards a complete safe path planning for robotic manipulators*, Intelligent Robots and Systems, 2010.
13. A. Ladikos, S. Benhimane, N. Navab: *Efficient Visual Hull Computation for Real-Time 3D Reconstruction using CUDA*, Computer Vision and Pattern Recognition Workshops, 2008.
14. M. E. Latoschik and H. Tramberend: *A scala-based actor-entity architecture for intelligent interactive simulations*, Software Engineering and Architectures for Realtime Interactive Systems, 2012.
15. A. Laurentini: *The Visual Hull Concept for Silhouette-Based Image Understanding*, Transactions on Pattern Analysis and Machine Intelligence, 1994.
16. S. M. Lavalle: *Rapidly-exploring random trees: A new tool for path planning*, Computer Science Department, Iowa State University, Technical Report, p. 98-11, 1998.
17. C. Lenz et al.: *Fusing multiple kinects to survey shared human-robot workspaces*, Technical Report TUM-I1214, Technische Universität München, 2012.
18. R. Marteniuk et al.: *Constraints on human arm movement trajectories*, Canadian Journal of Psychology, vol. 51, no. 3, pp. 365-378, 1987.
19. A. Ober-Gecks, M. Hänel, D. Henrich, T. Werner: *Fast multi-camera reconstruction and surveillance with human tracking and optimized camera*

configurations, International Symposium on Robotics, 2014.

20. A. Ober-Gecks, D. Henrich, M. Zwicker: *Efficient graphics processing unit-based voxel carving for surveillance*, Journal of Electronic Imaging, vol. 25, no. 4, 2016.
21. M. Quigley et al.: *ROS: an open-source Robot Operating System*, International Conference on Robotics and Automation, 2009.
22. H.J. Ralston et al.: *Energy-speed relation and optimal speed during level walking*, Internationale Zeitschrift fuer angewandte Physiologie einschliesslich Arbeitsphysiologie, vol. 17, no. 4, pp. 277-283, 1958.
23. J. Salvador, J. R. Casas: *Shape from Probability Maps with Image-Adapted Voxelization*, Workshop on Multi-camera and Multi-modal Sensor Fusion Algorithms and Application, 2008.
24. A. Saxena et al.: *RoboBrain: Large-Scale Knowledge Engine for Robots*, arXiv:1412.0691, 2014.
25. D. Stengel, T. Wiedemann, and B. Vogel-Heuser: *Efficient 3d voxel reconstruction of human shape within robotic work cells*, Mechatronics and Automation, 2012.
26. J. Ulmen, M. R. Cutkosky: *A robust, low-cost and low-noise artificial skin for human-friendly robots*, International Conference on Robotics and Automation, 2010.
27. G. Weiss: *Multiagent Systems: A Modern Approach to Distributed Artificial Intelligence*, ISBN 978-0262731317, The MIT Press, 2000.
28. J. Wen et al.: *Multi-agent based distributed control system for an intelligent robot*, Services Computing, 2004.
29. T. Werner, D. Henrich: *Efficient and Precise Multi-Camera Reconstruction*, International Conference on Distributed Smart Cameras, 2014.
30. T. Werner et al.: *ENACT: An Efficient and Extensible Entity-Actor Framework for Modular Robotics Software Components*, 47th International Symposium on Robotics, 2016.
31. K. M. Wurm et al.: *OctoMap: A Probabilistic, Flexible, and Compact 3D Map Representation for Robotic Systems*, International Conference on Robotics and Automation, 2010

Challenges of the Miniaturization in the Electronics Production on the example of 01005 components

Jens Niemann[1*], Stefan Härter[1], Christopher Kästle[1] and Jörg Franke[1]

[1]Institute for Factory Automation and Production System (FAPS),
University Erlangen-Nuremberg, Nuremberg, Germany
*jens.niemann@faps.fau.de

Abstract. The electronics production experiences a continuous trend towards miniaturization of components. Smaller components lead to a number of challenges in the Surface Mount Technology (SMT) process chain to ensure a stable production process. This paper provides an overview on the main SMT production processes and focuses on the stencil printing process, which is said to be an error-prone process and has the demand of further optimization and research. One important criteria for the solder paste printing stencils is the area ratio (AR), which is defined by the IPC-7525 as the area of the aperture opening to the area of the aperture walls. Especially in mixed assembly designs AR is becoming critical low for the miniaturized 01005 components. In this paper, the printing process for miniaturized components is investigated using specially designed area ratio test stencils and an active squeegee system. The design of these stencils includes values for the AR from 0.65 down to 0.45 with different aperture shapes, which include gradually adjusted in length and width, rectangles as well as circular apertures for reference. Furthermore, the orientation of the rectangular apertures in relation to the printing direction is regarded. To include the effect of the active squeegee system all tests were conducted with the system activated as well as deactivated.

Keywords SMT Production, Miniaturization, 01005 components, stencil printing

Introduction

In the area of Surface Mount Technology the miniaturization reflects an ongoing trend. The miniaturization leads to new technical requirements of the production processes and to highly integrated electronic systems. Especially the field of passive components (resistors, capacitors) component sizes of 01005 and 0201 (metric) with dimensions of 400 μm x 200 μm and 300 μm x 150 μm will be adressed and the trend goes towards further miniaturization. This leads to a high need of research to help the industry with the integration of these components in their product design and to enable them to establish stable manufacturing processes. Figure 1 provides an overview of today's component sizes used by the electronics industry.

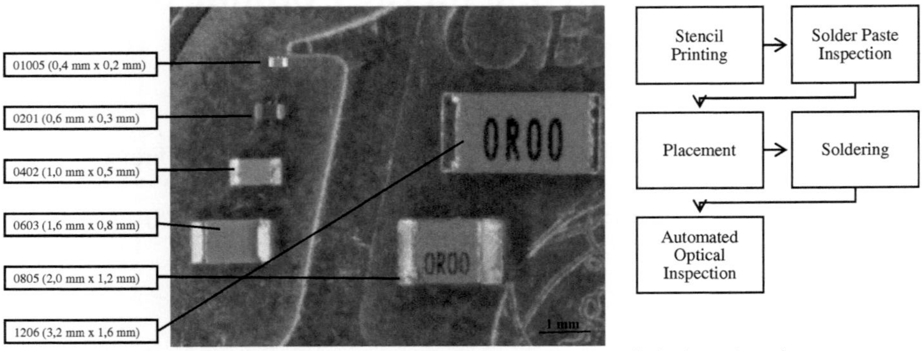

Fig. 1. different components sizes in the current SMT Production (left picture) and an schemantic picture of the SMT Process Chain(right picture).

This paper focusses on the SMT technology, the standard process in today's mass volume electronics production. In the beginning the study gives an overview of the SMT process chain and focusses afterwards on the stencil printing process, because it is the most error-prone process step, which causes nearly 70% [1–4] of all production failures. For the enhancement of the printing process there are two typical ways. On the one hand it is possible to change the process itself, e.g. by changing the solder paste material, the process parameters or the machine set-up, and on the other hand it is advisable to adjust the stencil layout to meet the new requirements [5, 6]. For the first optimization strategy the conducted experiments will evaluate the results of an active squeegee system, to show its influence on the printing performance. Furthermore, the stencil layout will also be subject of the investigation by testing different aperture forms and orientations.

The SMT Process Chain

The SMT Process Chain usually consist of three production and two inspection steps. It begins with the printing of solder paste, followed by the Solder Paste Inspection (SPI) to control quality parameters such as the so called transfer efficiency, which compares the printed solder paste volume with the volume of the stencil aperture. After the printing process the components are automatically placed into the solder deposits. The next step is the soldering of the printed circuit board (PCB) to establish a conductive and mechanically reliable connection. The final step is the Automated Optical Inspection (AOI) to check the correct position and further defects of all components. [7] The trend of miniaturization affects all production steps and in the following an introduction of the challenges will be given.

Stencil Printing The process of stencil printing is a widely spread process, which underlies several influencing factors. Especially the stencil thickness is becoming a critical factor concerning the continuous trend of miniaturization. Today's PCBs usually contain different component sizes. Thereby the required solder paste volume differs. Large components need a high volume of solder paste, which equals a thicker stencils and small components require only a small amount of solder paste volume, which can be achieved by a small stencil. It is also possible to reduce the volume by shrinking the aperture dimensions, which can lead to a reduced printing performance. The dimensions of the aperture are determined by the area ratio, which is defined by IPC-7525 as the area of aperture opening to the area of the aperture walls. Accordingly, the use of values for AR of at least 0.66 and above provides good printing results. For the overall PCB layout, a thicker stencil is used and the AR values for the miniaturized components drop below the limit of 0.66, which is critical to print and demands a perfectly designed stencil layout.

To ensure a high printing performance there are several ways to optimize the process. The stencil technology offers currently two broadly accepted solutions. One of them is the use of step stencils, where the stencil thickness is not constant, but varies to supply certain areas with a higher amount of solder paste. Second, the stencil surface can be retreated by using coatings to reduce the adhesion of the solder paste and the aperture walls. [8, 9]

Furthermore, there are different types of solder paste, which differ by the distribution size of the solder balls. A reduced mean solder ball size enables the printing process to print smaller apertures successfully, because the risk of the aperture to become clogged by solder balls is reduced. [6]

The most recent option to improve the printing process is the use of alternative squeegee technologies. One approach is the use of vibrating squeegee blades. The ultrasonic vibration of the blade leads to an improved behavior of the solder paste during the filling process of the aperture and thereby increases the printing performance of especially miniaturized apertures. [6, 8, 10]

Component Placement In general the machine capability of state-of-the art placement machines is sufficient to ensure a reliable placement process for miniaturized components. The processing of these components is sometimes limited by the component quality (e.g. delivery quality) and the way they are provided by the feeder. To process miniaturized components there is a need of precisely manufactured nozzles, which are able to pick-up the component correctly. This ability requires a specialized design to ensure long-term reliability of the nozzles for a robust placement process.

Soldering The challenges of the soldering is based on the PCB design, which is often a mixed-assembly board. Beside thermal influences of the PCB due to varying amount of copper, large components are placed next to miniaturized components, which have a different thermal behavior. Small components have a low thermal mass and can heat up faster, so these components would need a different temperature profile then large components, which have a higher thermal mass and need more time to reach the same temperature. Besides the heating up phase the behavior of cooling down is also different depending on the thermal mass, because the small components cool down faster than the large components. Both effects lead to a higher thermal stress for the miniaturized components. The strategy to reduce this thermal stress is an optimized temperature profile of the reflow oven, which is suitable with lowest thermal stress for miniaturized as well as for large components on the one side and to ensure a secured time above liquidus temperature for all interconnection parts. [11]

Experimental Setup

The following will present the experimental set-up of the conducted test on the printing process, which aim towards the evaluation of the active squeegee system and of the aperture form and orientation.

PCB Layout The used PCB is a black anodized aluminum plate with the dimensions of 160 x 100 x 1.5 mm. The used substrate represents nearly perfect conditions for printing tests and avoids most influences, which could be caused by the PCB. Furthermore, the black color enhances the contrast for the solder paste inspection machine and improves the measurement quality

Solder Paste The test aimed to achieve results that shows a comprehensive picture of the performance of an active squeegee system and the influence of the aperture shape. Therefore, four different kinds of no-clean SAC305 solder paste were used. They differ in terms of the solder paste type (type 4 and type 5), which represents different solder ball sizes (average size of solder balls for type 4 20-38 µm and for type 5 15-25 µm) and by manufacturer. Each manufacturer uses a different flux and sometimes a different mix ratio of flux and metal powder. The use of two manufacturers helps to investigate, if there is any influence of the flux. It is assumed that type 5 with the smaller solder balls achieves higher printing performances.

Stencils The stencils used were made of stainless steel foils with laser cut apertures and produced for attachment in a Vector Guard (VG260) master frame system. There was no stencil coating applied. In total the test were conducted with three different stencils. They differ in thickness and in the aperture shape, whereas the general structure remained identical. The stencils are structured in rows and columns. Each row represents a certain area ratio (0.45 – 0.65) and the columns stands for a different aperture shape. The shape includes a square and several different rectangles. Each rectangle is positioned twice on the stencil, meaning that one time the long side of the rectangle faces the squeegee and one time the small side. Thus, the influence of the aperture orientation can be checked. In Figure 2 all tested shapes are presented. The number represents the column and the dimensions of the rectangles in column 6/8, 5/9, 4/10, 3/11, 2/12 are always the same but the rectangles in columns 8-12 are rotated by 90°.

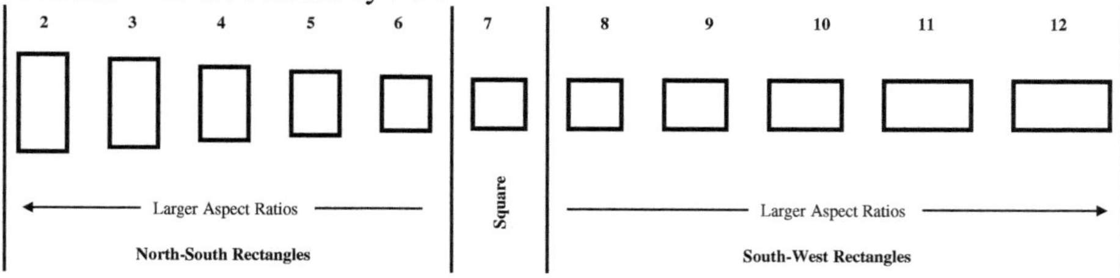

Fig. 2. Overview of all tested shapes.

Print Test Procedure All test were printed on an ASM DEK Horizon 03iX stencil printer and the solder paste was inspected by a Koh Young 3020T system with a 10 µm resolution. To select the testing configuration of the print parameters squeegee speed, squeegee pressure and the separation speed a pretest was conducted and the values were determined by a DoE optimization using the Minitab software program. To check, if the effect of the active squeegee system depends on the squeegee speed, two different speeds were tested. Table 1 shows the parameter configuration for the print test.

Table 1. Parameter Configuration of the conducted tests.

Paste Type	Squeegee Speed	Squeegee Pressure	Separation Speed
Type 4	Fast: 80 mm/s Slow: 30 mm/s	3 kg per 100 mm	2 mm/s
Type 5	Fast: 80 mm/s Slow: 30 mm/s	3 kg per 100 mm	2 mm/s

The test routine started with a preparation of four knead cycles, followed by an automatic under stencil wipe. Afterwards the test procedure was applied: ten boards at fast speed followed by ten boards printed of slow speed. The test print sequence was repeated twice, one time each with the active squeegee system activated/deactivated and using both paste types and brands for all three stencil thicknesses. The under stencil cleaner was programmed to perform a wet/vacuum/dry cycle and was applied after every fifth consecutive print. The complete DoE comprised 480 solder paste inspected boards with 3,208,804 solder pads in total.

Active Squeegee Systems

The main object in this experiment is the investigation of the DEK ProActiv system (squeegee length: 200 mm) as an example of an active squeegee system. In the first part the general result will be taken into consideration with the analysis of all measured data of the transfer efficiency. In further steps the result will be filtered by the criteria of squeegee speed and stencil thickness. The assumption is that the vibrations of the active squeegee system improve the general printing performance.

The general result is shown in Figure 3 and the transfer efficiency with the active squeegee activated and deactivated is compared. In this case the different aperture forms and the aperture orientation is not further investigated. The course of the graph reveals a positive effect of the active squeegee system on the transfer efficiency. The x-axis includes all tested area ratios and for every area ratio the activated squeegee system performs better than the prints, which were conducted with the system deactivated. Besides the transfer efficiency the standard deviation is an important factor for the process evaluation. Thereby the active squeegee system does not indicate any significant improvements of the printing process.

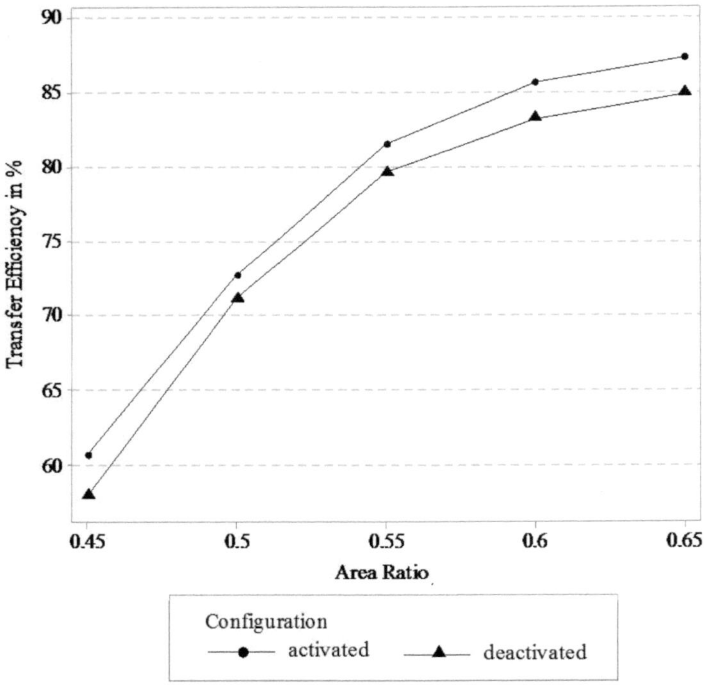

Fig. 3. Comparison of the DEK active squeegee system (activated/deactivated) depending on the area ratio.

Besides the general results, the analysis includes different parameters, especially the stencil thickness and the squeegee speed. Considering the stencil thickness an effect was determined, which proofed that a thicker stencil increases the performance of the active squeegee system and increases the transfer efficiency.

The test set-up was defined with two different squeegee speeds for each solder paste type. The assumption is that a slow speed leads to a higher performance, because the influencing period of the vibrations increases with a decreasing speed. The results indicate an increase of nearly 4 % on average of the transfer efficiency and a reduced standard deviation.

The Effects of the Aperture Design

The main goal of this experiment is to test what kind of influence is caused by the aperture shape and the orientation depending on the printing direction. In this study all results will be taken into consideration, meaning that any differences caused by e. g. the solder paste are not in the scope of this publication. The data will be evaluated by the criteria transfer efficiency (fig. 4 top) and standard deviation (fig. 4 bottom).

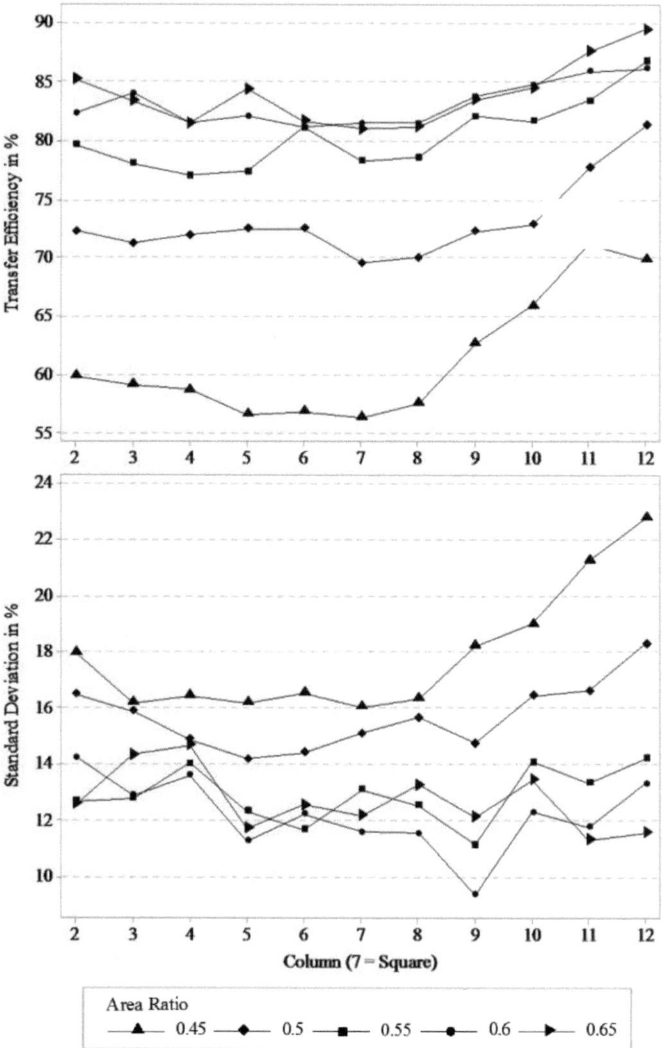

Fig. 4. Influence of aperture shape and orientation based on the transfer efficiency (top) and on the standard deviation (bottom).

In general, the square apertures achieve 8 % higher transfer efficiency than the circular structures. In the following, the variations of the rectangular apertures will be analyzed. In Figure 4 the overall view on the transfer efficiency and the standard deviation is visualized. The form is represented by the column number (Figure 2). The correctness and the plausibility of the data is proven by the trend of a rising transfer efficiency with increasing area ratios. Considering the influence of the aperture form and orientation the data shows a clear influence. The east-west orientated rectangles exhibit highest transfer efficiency (column 12). Starting from the square (column 7), which seems to achieve the lowest performance in nearly every graph, both rectangle orientations perform better than the square. Furthermore, the aspect ratio, which is rising towards the ends of the graphs, shows a possible correlation with the transfer efficiency. In addition to the aperture aspect ratio, the aperture orientation was investigated. There is evidence that the aperture orientation has an influence on the printing process. The apertures representing columns 2-6 have the small side of the rectangle facing the squeegee (i. e., north-south orientation) whereas the columns 8-12 are east-west oriented (i. e., the wide side faces the squeegee). The effect seems to depend on the area ratio. The lower area ratios of 0.45 and 0.5 show the clear tendency that there is an influence, especially at the edging rectangles. On average and compared to the square as a reference the transfer efficiency of the 0.45 area ratios raises about 3.5 % for north-south orientation and about 8.8 % for east-west oriented. This effect is decreasing, when all graphs are taken into consideration but throughout all graphs the trend can be recognized.

Further analysis on the solder paste and the stencil thickness reveal more influences. The solder paste type has an effect on the printing performance especially when narrow apertures are printed. In addition, it was investigated that thicker stencils due to their highest aspect ratios achieve higher transfer efficiencies.

Figure 4 (bottom) shows that the standard deviation of the area ratios of 0.45 and 0.5 reaches the highest values. A detailed analysis of the course of the graph shows an increase of the standard deviation beginning from the square towards the ends of the graph. As with the transfer efficiency this trend is remains recognizable throughout all area ratios.

In conclusion the aperture form as well as the aperture orientation have an influence on the transfer efficiency and the standard deviation. A rectangular shape accounts for a better transfer efficiency especially in the east-west orientation but it also leads to a higher standard deviation, which might limit the process capability.

Discussion

The experiments proofed that these two approaches (usage of an active squeegee system and the optimized stencil layout) to enhance the printing performance of miniaturized structures are valid. In the following the results will be discussed.

The positive effects of the active squeegee system are based on the improvement of the sub processes of the printing process. The vibration improves the so called rolling (which is the movement of the solder paste during the printing process) of the solder paste by reducing the friction between the solder balls and the surface of the stencil and the squeegee blade. Furthermore, it seems to influence the filling process of the apertures and the alignment of the solder balls in the aperture. The vibration might improve the solder balls array and thereby enable the printing of miniaturized structures. The last improvement of the active squeegee system is in the last step of the printing process, which is the separation of the PCB and the stencil. Especially for miniaturized structures there is a high adherence between the aperture wall and the solder paste. While separating the solder paste sticks to the wall and thereby reduces the transfer efficiency. The vibration again helps to reduce the friction and improve the printing process.

The results of the experiments comparing the aperture form show the influence on the stencil printing process especially for miniaturized structures with area ratios below 0.6. The upcoming discussion introduces some ideas for the integration of the results in present standard rules.

The first insight of the experiment shows that a rectangle shape leads to better results than the square. Combined with the analysis of the influence of the stencil thickness it is known that the effect might depend on the aperture size. All stencils have the same area ratio, due to the calculation of the area ratio it also means that the absolute size of each aperture is not comparable and that the absolute size might also be another criterion, which should be integrated in the stencil design. Continuing with the aperture orientation analysis it was found out that there are differences between the performance of the east-west and the north-south directions. The test data has revealed the orientation and form of the aperture influences the printing performance, particularly at miniaturized structure where unexpectedly high average paste transfer emerged.

Outlook

Miniaturization will be a continuous topic for the electronics production. The approaches shown in this publication are based on one comprehensive experiment. For the active squeegee system, it is required to find out more about the exact principle of operation and to further test different frequencies to optimize the effect of the ultrasonic vibrations.

The test of the aperture form and orientation lead to the question, if the current rules and standard defined by the IPC need to be more precise and extended to meet the requirements of the miniaturization

.

Acknowledgement

The authors would like to thank Mr. J. Schake and Mr. M. Whitmore (both DEK/ASM) for supplying machinery and materials under investigation as well as their advice. The authors are deeply grateful for this support and are responsible for the content of this publication.

References

[1] K. VIJAY: *Miniaturization - Solder Paste Attributes for Maximizing the Print & Reflow Manufacturing Process Window*, 18th European Microelectronics and Packaging, Brighton. In: *IMAPS-Europe*, S. 26–29

[2] C. COMBET, M.-M. CHANG: *01005 Assembly, the AOI route to optimizing yield*. In: *Vi TECHNOLOGY* Jul. 2009

[3] R. S. CLOUTHIER: *The complete solder paste printing processes*. In: *SMT Magazine* vol. 13 (1999), S. 6–8

[4] C. H. MANGIN: *Where quality is lost on SMT boards*. In: *Circuit Assembly* no. 2 (1991), S. 63–64

[5] M. WHITMORE, J. D. SCHAKE, C. ASHMORE: *The impact of stencil aperture design for next generation ultra-fine pitch printing*. In: *Proceedings of the 35th International Electronics Manufacturing Technology Conference* (2012), S. 1–7

[6] S. HÄRTER ; J. NIEMANN, J. FRANKE, M. WHITMORE, J. D. SCHAKE: *The Effect of area shape and area ratio on solder paste printing performance*. In: *Proceedings Surface Mount Technology Association International Conference (SMTAI)* Rosemont, IL (2016), S. 613–620

[7] M. RÖSCH: *Potenziale und Strategien zur Optimierungdes Schablonendruckprozesses in der Elektronikproduktion*. Bamberg : Meisenbach, 2011 (Fertigungstechnik 221)

[8] S. HÄRTER, C. LÄNTZSCH, J. FRANKE: *Fundamental study on a secure printing process using NanoWork stencils for 01005 components*. In: *Proceedings Surface Mount Technology Association International Conference (SMTAI)* Rosemont, IL (2014), S. 878–884

[9] M. RÖSCH ; J. FRANKE ; C. LÄNTZSCH ; G. KLEEMANN: *Characteristics and Potentials of nana-coated stencils for stencil printing optimization*. In: *Universität Erlangen-Nürnberg* FAPS (2010)

[10] S. HÄRTER: *Evaluation of the Stencil Printing for Highly Minia-turized SMT Components with 03015mm in Size*. In: *Proceedings Surface Mount Technology Association International Conference (SMTAI)* Rosemont, IL (2015), S. 756–764

[11] F. SCHÜßLER, D. KOZIC, J. FRANKE: *Influences on the reflow soldering process by components with specific thermal properties*. In: *Circuit World* Vol. 35 (2009), S. 35–42

Boundary Conditions in Handling of Microspheres Induced by Shape Deviation Constraints

Finn Meiners[1*], Sebastian Hogreve[1], Kirsten Tracht[1]

[1]Bremen Institute for Mechanical Engineering (bime), University of Bremen, Badgasteiner Str. 1, 28359 Bremen, Germany
meiners@bime.de

Abstract. For speeding up the development of new metallic materials, the collaborative research centre SFB1232 'Farbige Zustände' researches a high-throughput method for material testing. This approach aims at generating metal spheres smaller than 1 mm in diameter, and applying different thermal and mechanical treatments on them. Afterwards the samples are inspected in different processes to gain information about the material characteristics. To move the samples through this system, a handling system is needed, which can provide the spheres in different handling modes for different processes. This article describes the identification of boundary conditions for a handling system, resulting from the geometric characteristics of the microspheres.

Keywords: Handling, Measurement, Methodology

1 Introduction

Conventional development of structural material is a time consuming process based on standardised samples and testing methods. Each test allows the determination of directly comparable qualities of the analysed material. The collaborative research centre SFB1232 'Farbige Zustände' researches a new high-throughput method for material testing which will increase the speed of the material development drastically. The aim of the collaboration is to build this high-throughput system, which analyses microscopic material samples with specifically developed test methods, allowing indirect conclusions on the standardised material properties [1]. In the past, high-throughput systems have been applied in the pharmaceutical industry for drug development and the development of new materials for catalytic converters [2, 3]. Usually these high-throughput systems do not use discrete samples but liquids, powders or thin-films which can be easily automatically applied to each single test [4].

In order to achieve the material tests in high-throughput, a handling system is required, which can process a high number of samples in a very short time. The high diversity of the different processes e.g. in the domain of mechanical and thermal treatment causes the necessity of a handling system, which can provide micro samples in different handling modes for different purposes. Depending on the requirements of the respective treatment or testing process, the samples have to be provided for example in bulk form, separated in trays or fixed on a carrier plate, either in whole or ground flat, see figure 1. For the fixation there are several methods, for example vacuum clamping and embedding in adhesives or low melting alloys. As a test sample takes its individual route through the treatment and inspection process, it may have to be transferred into a different handling mode, either because of batch creation for efficient workpiece flow, or because of different requirements of the following process. This creates the demand for a fast handling system, which can move single specimen from one mode to another. This system has to meet multiple requirements, which can be separated in demands on the high-throughput, requirements resulting from different treatment and inspection processes and boundary conditions resulting from the characteristics of the microspheres. The aim of this research is to identify the characteristics of the microspheres.

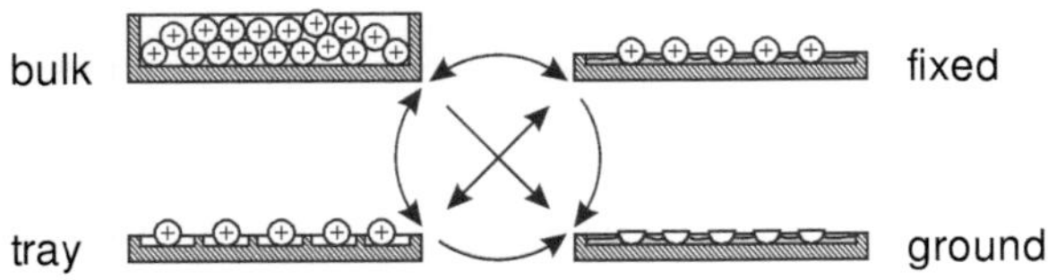

Fig. 1. Different handling modes and transfers

2 Handling of Microparts

The handling of micro parts is related to specific physical effects, which are usually irrelevant in the macro scale. With decreasing size of the parts, the mass decreases by the third power, while the surface decreases only by the second power. This causes special effects depending on the surface to gain importance, like surface tension, Van der Waals force, or electrostatic force [5]. Also micro parts can float on top of liquids because the surface tension is stronger than the gravitational force. From these properties result restrictions for handling methods, for example the separation of a micro part from a gripper might not be guaranteed because of adhesion forces. Consequently, many handling methods used in the macro scale are not usable for micro parts. [6]

Usable gripping principles depend heavily on how small the parts are. Until a specific point, friction-based gripping, formclosed-gripping, vacuum gripping and magnetic gripping are applicable [7]. Other principles specifically developed for micro parts are based on the force relationship in the micro-domain.

For researching the boundary conditions, which different gripping and clamping methods have to satisfy, it is necessary to know the spheres' geometric properties like their diameter and surface characteristics. Of particular importance are the minimum and maximum values rather than only the mean characteristics, because the extremes define the constraints for applicable handling systems.

3 Analysis of Characteristics of Microspheres

To gather the information, preliminary test samples are inspected, which are made by single droplet solidification. This process creates small droplets in an on-demand droplet generator from molten metal. The process can be controlled by changing the valve opening time and pressure of the gas, which is applied to push the melt through a fine nozzle. The specimen fall through a tube filled with inert gas where they solidify, until they reach a beaker with a quenchant. [8] The copper microspheres, which are object of the investigation, are divided into three different batches, created with different pressures and valve opening times, see table 1.

Table 1. Process parameters of tested microspheres

Batch	Pressure (mbar)	Valve opening (ms)	Pressure curve
A	200	3	A
B	230	3.5	B
C	320	2.5	C

3.1 Sampling

Taking a randomly created subset from a batch with differently sized microspheres is a challenge in itself. Simply pouring the microspheres from one bin to another can result in a bias, caused by segregation processes. To overcome this problem, a single line of microspheres is laid out, and each n-th is picked for further investigation, see figure 2, left. These samples are placed on double-sided adhesive tape on top of a microscope slide for further inspection, see figure 2, right.

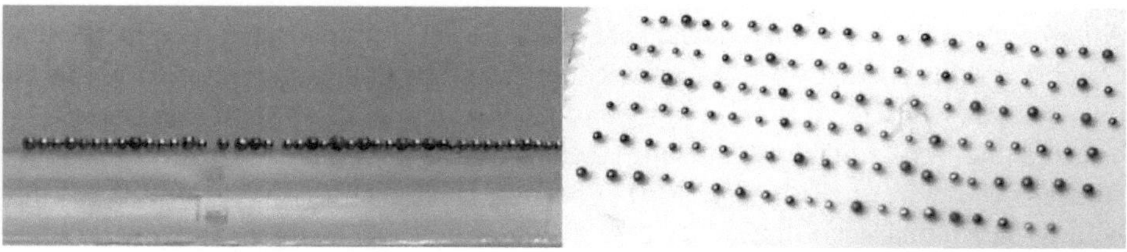

Fig. 2. Microspheres lined up for sampling (left) and on carrier plate (right)

3.2 Measuring

The specimens are measured with the laser scanning microscope, which creates a 2.5 dimensional heightmap of the sample by measuring the focus height of each pixel in small steps for the whole image. The resulting image has a size of 2048 times 1536 pixel. Two different lenses are used for the measurements. A 50x lens with its numeric aperture of 0.95 is able to resolve fine details, but with a focus distance of only 350 μm there is a risk of colliding with larger spheres in an automatic process, when the lens is lowered too much. The 20x lens has a focus distance of 3.1 mm which allows a safe operation with different sized spheres. Furthermore, this lens is capable of recording each sphere in a single step without stitching several images together, which would take more time and could lead to additional errors.

The resulting heightmap contains roughly a third of the sphere. Of course the laser beam cannot reach anything below the equator of the sphere, but even before this line, height information is missing. This happens because most of the light, which reaches the sphere surface in a very flat angle, is not reflected back to the sensor. To compensate this effect partly, the whole scanning process is repeated a second time with a higher luminous intensity. Especially the 50x lens creates noisy data in this area, because of the high numeric aperture, respectively the big opening angle of the lens. This causes the sphere itself to cast a shadow on the measured point for a part of the light beam, see figure 3.

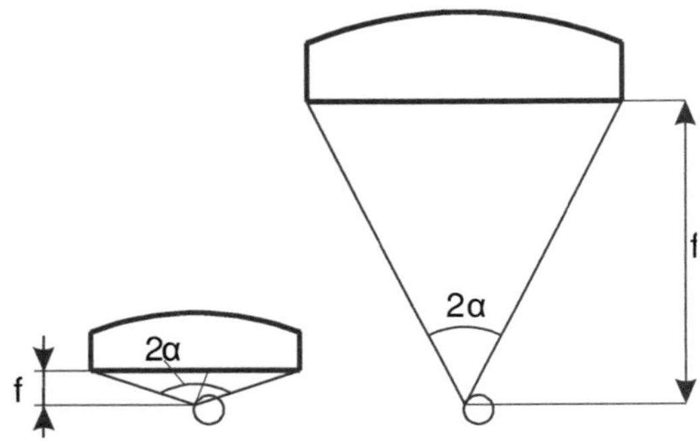

Fig. 3. Focus distance f of lens with large (left) and small (right) numeric aperture

3.3 Processing Data

After creating the heightmap of the spheres, there are multiple ways on how the data could be evaluated. They are either based on functions in the analysis software of the microscope (VK-H1XAD) or on functions developed in MatLab for this purpose.
To find the best suitable evaluation method, they were all tested with reference spheres. These are steel (X105CrMo17) bearing balls with a nominal size of 0.3 mm with a G10 rating according to ISO 3290 [9], which means their radius is within a tolerance of 0.25 µm. The results are listed in table 2.

The simplest approach is to take the difference between the highest point of the sphere and the sample carrier as the spheres diameter. Due to the soft adhesive tape, into which the sphere can sink, this measurement is not accurate.

The automatic function for generating stacked focus images in the microscope's software is much faster, than the laser scanning process, but it is not always clear enough, where to draw the border between the sphere and the background. Furthermore it provides only 2D information, as depicted in figure 4.

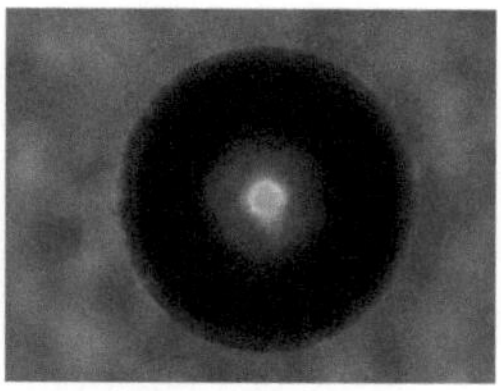

Fig. 4. Stacked focus image

The integrated functions for sphere measurement in the microscope's analysis software give accurate results. The main drawback of this process is the lack of automation. For each sphere, the measurement area has to be marked by the user to separate it from the background. This is not only a disadvantage because the process is time consuming and thus limits the number of samples, but the measurement result varies slightly with the choice of the marked area and by this it is not repeatable.

To overcome this challenge, an algorithm has been implemented in MatLab, to automatically process large numbers of spheres under repeatable conditions.

Table 2. Comparison of different measurement and evaluation methods for reference spheres

Method	Highest point to plate	2D image with stacked focus	Keyence sphere measurement	Keyence sphere measurement	MatLab least square
Lens	x20	x20	x20	x50	x20
Ball 1	285 μm	305.5 μm	300.45 μm	302.02 μm	300.14 μm
Ball 2	289 μm	304.2 μm	300.71 μm	301.92 μm	299.80 μm
Ball 3	291 μm	303.8 μm	299.43 μm	301.35 μm	298.42 μm
Mean deviation	11.7 μm	4.5 μm	0.57 μm	1.76 μm	0.64 μm

The resulting data is transformed from the proprietary format to a CSV file, which can be imported to MatLab. The data of the sample is separated from the background and the overshadowed area by only using points with a height above 3/4 of the maximum height of the sphere. A least square algorithm [10] is used to find the best fitting equivalent full sphere for each microsphere, as shown in figure 5.

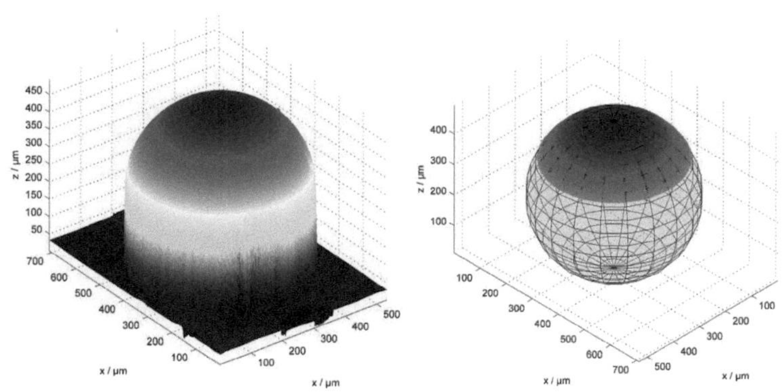

Fig. 5. Left: raw heightmap visualised in MatLab. Right: cropped heightmap with fitted sphere

4 Results

4.1 Size

For each batch, 120 samples are placed on the carrier. Some of the measurements are filtered out, because they show disturbances not belonging to the actual spheres. The resulting sample sizes are: batch A: 120, batch B: 116, batch C: 109. The mean values are $\bar{x}_A$= 227 µm, $\bar{x}_B$= 201 µm, $\bar{x}_C$= 216 µm and the standard deviations σ_A=29.9 µm, σ_B=28.5 µm and σ_C=34.3 µm. The histogram of the sphere sizes is shown in figure 6. It reveals a wide spread of the sphere diameters, e.g. the sizes in batch C are in a Range from around 150 µm to 300 µm. This means, the biggest spheres are double the size in diameter, four times the surface and eight times the weight.

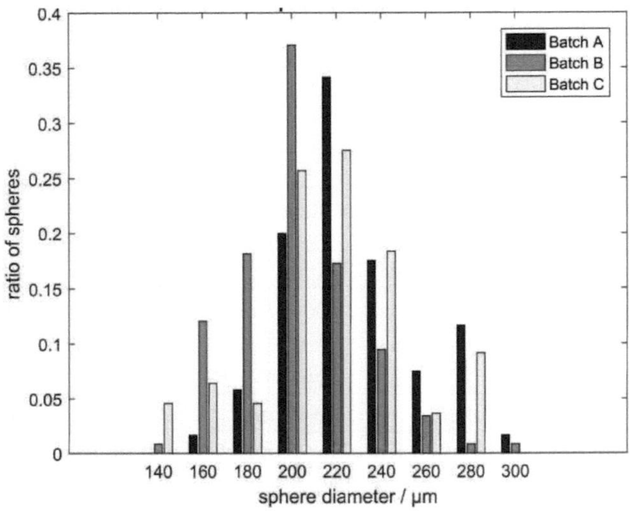

Fig. 6. Distribution of sample diameters for different batches

4.2 Surface

Additionally to the size, the surface of the spheres was analysed. A first glance on the samples (see figure 7) shows a wide variance of colours, from golden to orange to brown and grey, which is correlated to the size of the spheres. While the smaller spheres show a high saturation, the bigger they get, the more their colour turns to grey and they also lack the shine of the smaller ones. On a closer look with the microscope, most of the spheres are not uniformly coloured. The more reddish spheres show a kind of leopard pattern of yellow, purple and white spots. Additionally to the colour, there are different surface structure types. While the golden spheres show a smooth, consistent texture, the grey spheres show a mosaic like structure resembling crocodile leather, with small, around 0.5 µm deep groves between smooth plateaus, see figure 7. These are the dendrites caused by crystal growth during the solidification.

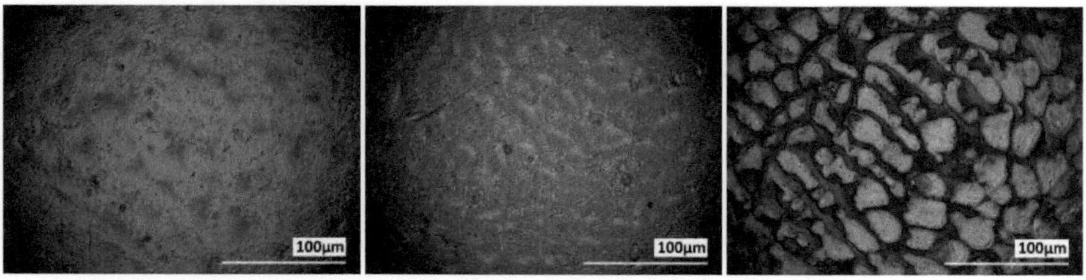

Fig. 7. Different sphere surface structures and colours

Many of the spheres show irregular impurities on their surface. These are small, confined areas, which stick out of the otherwise smooth surface. Judging by the colour, and their texture, they seem to be of the same material as the spheres, and not later added particles.

5 Discussion and Conclusion

The results of the measurements reveal boundary conditions, which a handling system has to comply with. The huge spread of sphere sizes requires a handling method, which can tolerate the wide range of diameters from 150 µm to 300 µm. This does not only limit positive fit connections, but every method, where close tolerances have to be met. If a suction gripper is used, undersized spheres might get sucked into the gripper, or get stuck in it. On the other hand, an oversized part might be too heavy to be lifted. In general, every gripping method which relies on the fine balance between surface forces for gripping, and gravitational forces for releasing, is limited in the range of the micro part's sizes.

A suitable clamping method has to be adaptive, to cover these wide tolerances. This can be accomplished by liquid fixation mediums, as with embedding the spheres in low melting alloys (form and force fit) or adhesive bonds, see figure 8.

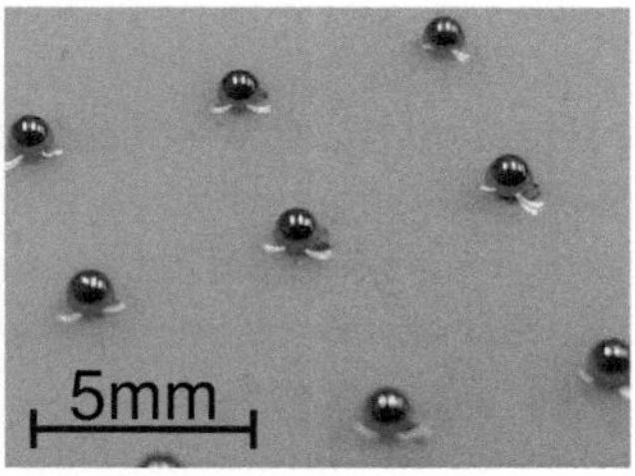

Fig. 8. Microspheres, embedded in epoxy for fixation

Another consequence of the widespread diameters arises, when the spheres are to be embedded to ensure a form fit. For this, the embedding medium has to reach over the equator to enclose the widest part of it. With a height of the embedding medium over half of the largest sphere, the smallest one would be covered completely, see figure 9.

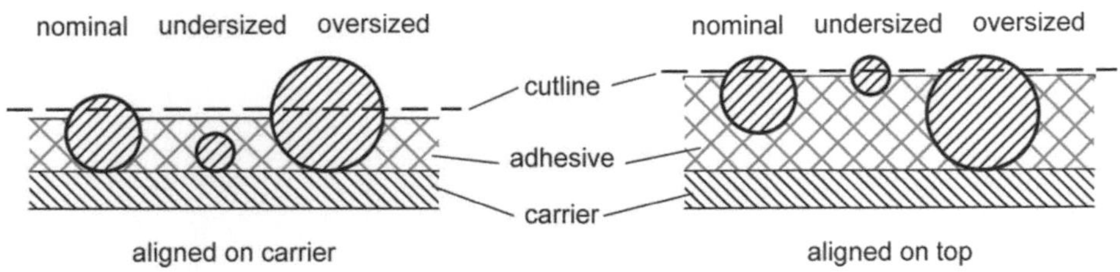

Fig. 9. Top and Bottom aligned spheres

For some of the measuring processes, the spheres have to be ground down to ensure an even surface. Here again, the different sizes would cause complications, as the smallest spheres could not be ground down, without removing more than half of the bigger sphere from its fixation, which would result in losing the form closure.

Both of these challenges could be overcome by aligning the samples at the top, e.g. by holding them in a fixture, until the embedding medium has solidified. But this would come at the disadvantage of a gap between the carrier and the sphere. This gap could distort stress and hardness tests, when not the sphere, but the embedding medium deforms under the applied load.

The above described conflictive boundary conditions can only be met by providing the spheres in different modes. These depend on whether the corresponding process requires top or bottom aligned spheres.

An alternative to adapting the handling process to the spheres would be to sort out spheres above and below acceptable limits by sieving. Likewise a colour based sorting process could be used, e.g. if a layer of oxide on the sphere prohibits a good bonding with adhesives or low melting alloys.

6 Limitations

Even though the sampling of the spheres has been carried out with great care to avoid biased results, it must be pointed out, that a first separation process has been carried out by their producer.

The tested samples do not represent the final spheres, as they are still work in progress. It is a snapshot of the current status, which is used to develop a decision process which can be applied to any later provided set of samples. Not only the material composition of the spheres will change, depending on which material has to be tested, but also the spread in size is expected to be smaller.

Additionally not all boundary conditions which are resulting from the different processes for structural change and measurement are known yet.

Acknowledgements

Financial support of subproject P04 ‘High Throughput Handling of Micro Samples’ of the Collaborative Research Centre SFB1232 ‘Farbige Zustände’ by the German Research Foundation (DFG) is gratefully acknowledged.

References

[1] Drechsler, R., Eggersglüß, S., Ellendt, N., Huhn, S., Mädler, L.: Exploring Superior Structural Materials Using Multi-Objective Optimization and Formal Techniques In: 6th IEEE International Symposium on Embedded Computing & System Design (ISED). International Symposium on Electronic System Design (ISED-06), December 15-17, Patna, India (2016)
[2] Dogan, C.: Optisches Hochdurchsatz-Screening nach sauerstoffreduzierenden Elektrodenkatalysatoren zur Anwendung in Brennstoffzellen, Dissertation Universität des Saarlandes (2014)
[3] Weiß, T.: Entwicklung neuer heterogener Katalysatoren zur Spaltung von Methanol in Wasserstoff und Kohlenmonoxid mittels Hochdurchsatz-Methoden, Dissertation Universität des Saarlands (2008)
[4] Baudry, S.: Einsatz kombinatorischer Hochdurchsatz-Methoden zur Entwicklung bleifreier Piezokeramiken, Dissertation Universität Karlsruhe (2010)
[5] Fearing, R.: Survey of sticking effects for micro parts handling. In: Proceedings of the IEEE/RSJ International Conference on Intelligent Robots and Systems, vol. 2, pp. 212–217 (1995)
[6] Fantoni, G., Santochi, M., Dini, G., Tracht, K., Scholz-Reiter, B., Fleischer, J., Lien, T.K., Seliger, G., Reinhart, G., Franke, J., Hansen, H.N., Verl, A.: Grasping devices and methods in automated production processes. In: CIRP Annals - Manufacturing Technology, vol. 63, pp. 679–701 (2014). doi:10.1016/j.cirp.2014.05.006
[7] Sánchez, A.J. Handling for Micro-manufacturing. In: Micromanufacturing Engineering and Technology, pp. 637–659 (2015). doi: 10.1016/B978-0-323-31149-6.00027-X
[8] Ellendt, N., Ciftci, N., Goodreau, C., Uhlenwinkel, V., Mädler, L.: Solidification of single droplets under combined cooling conditions. In: IOP Conference Series: Materials Science and Engineering, vol. 117 (2016) doi:10.1088/1757-899X/117/1/012057
[9] ISO 3290-1 Rolling bearings - Balls - Part 1: Steel balls, Beuth (2014)
[10] Jennings, A.: (2011). Sphere Fit (least squared) (https://de.mathworks.com/matlabcentral/fileexchange/34129), MATLAB Central File Exchange. November 11, (2016)

Continuous Digitalization of Rotationally Symmetrical Components With a Lateral Scanning White Light Interferometer

Sönke Bahr[1], Marc-André Otto[1], Tomas Domaschke[2], and Thorsten Schüppstuhl[2]

[1] VMT Vision Machine Technic Bildverarbeitungssysteme GmbH
Mallaustr. 50-56, 68219 Mannheim,
soenke.bahr@vmt-systems.com, http://vmt-technology.com
[2] Technische Universität Hamburg-Harburg
Institut für Flugzeug-Produktionstechnik, Denickestraße 17, 21073 Hamburg,
ifpt@tu-harburg.de, http://ifpt-tuhh.de/

Abstract. In the field of aircraft maintenance the demand for automation and increased reproducibility is currently growing. With the inspection of aircraft combustion chambers it has been shown in a previous research project that this can also be achieved for complex tasks. A fully automated process was developed with a robot-guided white light interferometer, which digitizes the entire combustion chamber of an aircraft engine in high resolution. Furthermore, damages such as cracks and breakouts are detected automatically [1,8]. In addition to a standard industrial robot a turning table is used to utilize the rotational symmetry of the part. Due to the special procedure the part must stop at each of the more than 50.000 measurements. This procedure is time-consuming and induces unwanted vibrations into the inspection system. In this proceeding a new approach is presented that uses the same sensor technology continuously. This leads to a significantly faster process and a noticeable reduction of vibrations. On the other hand, considerable adaptations are necessary in the field of measurement technology and handling. These topics will be examined in more detail below.

Keywords: lateral scanning white light interferometer, robot-guided, on-line adaptation, airplane MRO, 3D data processing

1 Introduction

Currently the knowledge-based fluorescent penetrant inspection method (FPI) is a widely used method in the aviation maintenance, repair and overhaul (MRO) industry. This completely manually executed inspection process for finding small cracks is time intensive, uses chemical fluids and is highly dependent on the human factor. Therefore, a new optical non-destructive testing (NDT) method for rotational symmetrical components (RSC) was developed in a research project initiated by Lufthansa Technik AG [3]. The results of this project have been presented by Domaschke et al. [1] and Otto et al. [8]. In this approach a white light interferometer (WLI) is handled by an industrial robot. The RSC is installed on and consequently handled by a turning table. For the whole inspection of the RSC 50.000 single measurements are required. These single measurements are merged to patches consisting of e.g. 7 times 7 scans to extend the field of view (FOV). The crack detection is performed on these patches and has shown that cracks with lengths as low as 1 mm can be found. Figure 1 shows the overall system setup.

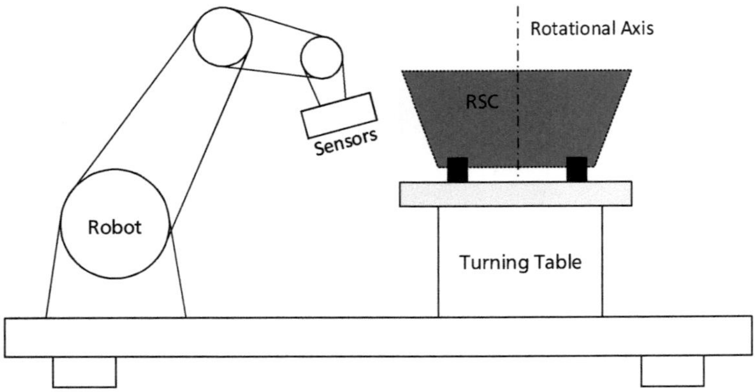

Fig. 1. Overall setup with robot, sensors, RSC and turning table

The RSC is only approximately rotationally symmetrical. It has a height of approx. 250 mm and a diameter of about 400 mm. In reality the part is deformed and cannot be perfectly centered on the rotational axis. Because of this in combination with the WLI's low measurement range, a laser triangulation sensor (LTS) is used in addition to digitize the RSC in a first step. With this data an adaptation of the WLI's measurement position to the surface can be achieved. In Chapter 2 the measuring procedure of this approach is described in detail.

2 Current approach

The WLI used in the current approach is constructed similarly to a Michelson-interferometer [2]. Figure 2 shows the optical setup of the WLI. The light beam emitted from the LED (a) is separated into two paths by the beam splitter (b). The reference path (green) is reflected by the reference mirror (d) and on the way back it is directed by the beam splitter to the camera (e). The measuring path (red) is reflected by the surface (c) itself and will pass by to the camera directly. Among the beam splitter and the camera the interference occurs between the optical measurement path and the reference path. The emitted white light has a broad wave length spectrum which results in a very short coherence length. The distance in which the reflection from the surface results to the highest constructive interference is called the coherence plane.

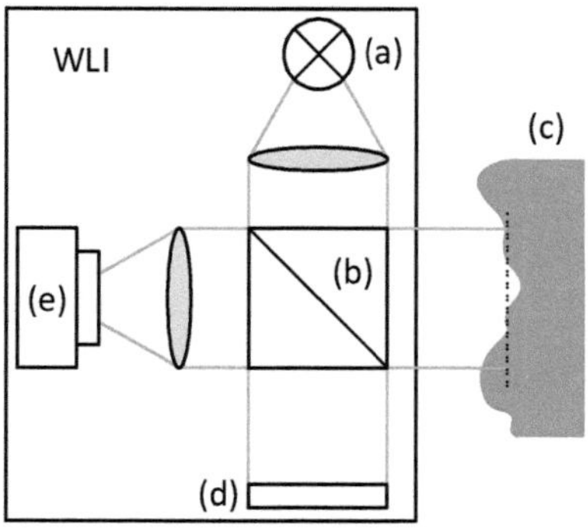

(a) Principle design

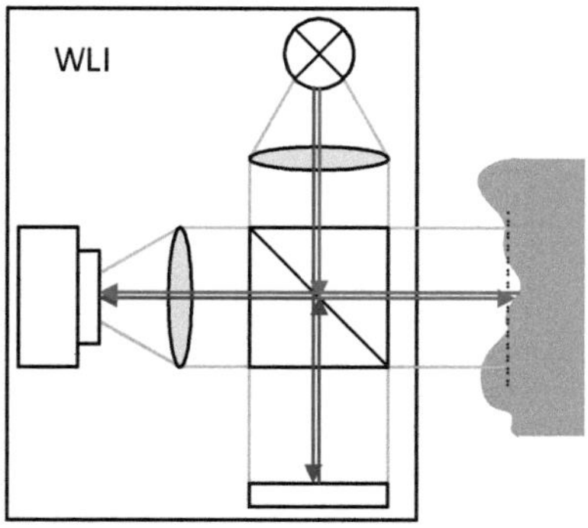

(b) Optical paths

Fig. 2. Optical setup of a WLI, (a) light source, (b) beam splitter, (c) surface, (d) reference mirror, (e) camera, (red) measurement path, (green) reference path

For the measurement the WLI is moved by the robot to a static position. Between the sensor and the robot flange a linear axis is mounted. With this axis the interferometer and with it the coherence plane will be moved vertically towards the surface. This vertical scanning interferometer (VSI) captures a defined number of frames in each measurement. The measured volume is represented by theses frames. The intensity for each pixel results from the interference between the optical measurement path and the reference path.
During exposure the intensity will be higher in areas, where the surface is near the coherence plane. In the post-processing for each pixel the captured correlogramm is examined. Thereby, a demodulation is performed and the frame is determined in which the highest intensity was measured. Afterwards the distance for this pixel is calculated with a weighted mean over the neighboring frames. In figure 3 the movement direction and the captured frames are presented simplified in a two dimensional abstraction. The red arrow illustrates the vector through the associated pixels.
After this procedure the rotational axis turns the part to the next position. Before the next measurement can be executed the system waits until the induced vibrations have subsided. In this process the part is measured in a ring-wise fashion. The robot is only moved after a ring was fully scanned, in order to reduce vibrations of the interferometer during measurement. The described approach can digitize the part with a high lateral resolution of 20 µm and a good measurement coverage.
Because of the large proportion of time, in which no frames are collected between two measurements, this approach is very time intensive. Furthermore, the jerky movement induces unwanted vibrations. In chapter 3 a novel approach will be proposed, which can compensate these disadvantages.

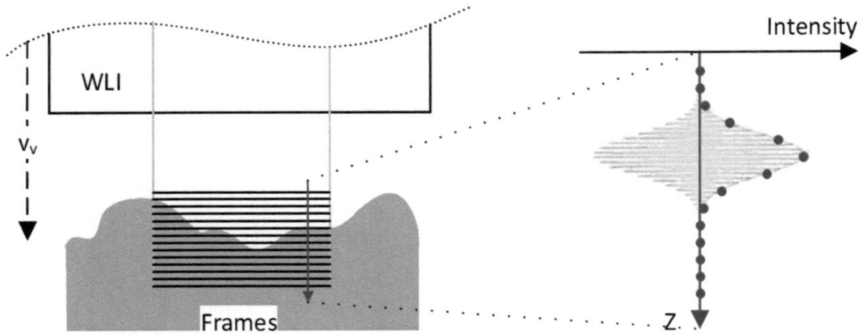

Fig. 3. Left: movement direction and captured frames for VSI, right: correlogram for one pixel

3 Novel approach

In addition to the use of the WLI in the manner described above, Olszak introduced a new method in 2000 [7]. In this novel approach the sensor is oriented at a small angle to the surface. The movement is lateral to the surface instead of vertical like with a VSI. Therefore, it is called lateral scanning interferometer (LSI). Because of the skewness, the velocity still has a component vertical to the WLI (see figure 4). The optical setup of the interferometer stays the same, but the position of frames relative to each other changes. In figure 4 these frame-positions are shown in global and in sensor coordinates. The red arrow clarifies, that the frames have to be run through diagonally instead of vertically.

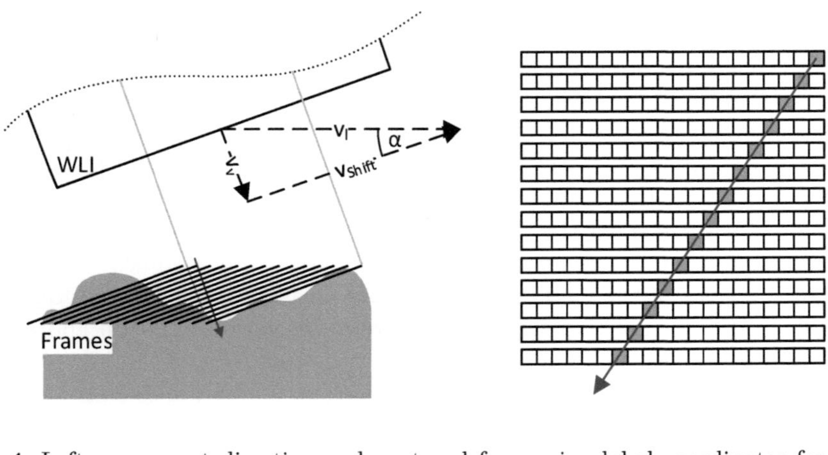

Fig. 4. Left: movement direction and captured frames in global coordinates for LSI, right: frames in sensor coordinates

Therefore, a correlogram can be extracted by shifting the regarded pixel position for each frame. The lateral velocity v_l should be chosen in a manner that the shift Δx is an integer multiple n_{Shift} of the pixel size s_{Pixel}. This pixel size is defined here as the distance between two neighboring points in the coherence plane and is dependent on the optical setup and thereby on the optical magnification. With the skew angle α the equation for the shift results as:

$$\Delta x = \frac{n_{Shift} \cdot s_{Pixel}}{\cos \alpha} \tag{1}$$

The lateral velocity results from the shift and the frame rate f_{Frame}:

$$v_l = f_{Frame} \cdot \Delta x = \frac{f_{Frame} \cdot n_{Shift} \cdot s_{Pixel}}{\cos \alpha} \tag{2}$$

The frame rate is chosen by the corresponding sensor parameters which will not be presented in this paper. For further information on the interdependency of these parameters refer to [4] and [2]. In the case of the RSC the lateral displacement is achieved by continuous rotational movement of the component past the static sensor. Thus the angular velocity ω has to be chosen so that the following equation (which depends on the radius r) is fulfilled:

$$\omega = \frac{v_l}{r} = \frac{f_{Frame} \cdot n_{Shift} \cdot s_{Pixel}}{r \cdot \cos \alpha} \tag{3}$$

4 Implementation

First practical experiments validated that the outlined approach works with some restrictions. In figure 5 a first result is shown on the left side. The image on the right side shows the same area measured with the VSI.

(a) Amplitude: LSI (b) A: VSI

(c) Distance: LSI (d) D: VSI

Fig. 5. First results with LSI compared to VSI

For both results the adaptation values of the VSI method were used as the measurement position. The data measured with the VSI fits in the vertical measurement range of approx. 3.5 mm. The LSI measurement illustrates that the data on the right as well as on the left side of the image are out of measurement range. Therefore, one of the main difficulties of this new approach is the limited measurement depth.

4.1 Measurement Depth

The measurement depth z_{Range} depends on the skew α of the coherence plane to the movement direction:

$$z_{Range} = y_{Range} \cdot \sin\alpha \tag{4}$$

The measurement range in y-direction y_{Range} results from the pixel count in that direction n_{Pixel} and the pixel size:

$$y_{Range} = n_{Pixel} \cdot s_{Pixel} \tag{5}$$

With a skew angle of 5° and 300 pixels with a size of 20 µm the resulting measurement depth is:

$$z_{Range} = n_{Pixel} \cdot s_{Pixel} \cdot \sin(\alpha) = 300 \cdot 20\,\mu\text{m} \cdot \sin(5^\circ) = 0.523\,\text{mm} \tag{6}$$

The sensor has to follow the surface. Otherwise parts of the surface would not be within the measurement range. This results in higher requirements for the handling and the post-processing technology. This adaptation can be achieved with the previously described approach. Off-line data can be generated by a sensor with a larger measurement range.

To reduce the total process time even further, the first digitalization step should be discarded. This could be solved through the implementation of an on-line control loop. There are two possible procedures: firstly, an LTS like the one for the off-line adaptation can be positioned to measure the surface right before the interferometer measures it. Alternatively the interferometer data itself can be used for the control loop. This is only possible if the surface is strictly continuous.

In all cases, in which the distance to the surface changes during the measurement, the corresponding frames have to be moved relative to one another. In a first step only an offset in z-direction of the sensor could be sufficient. For the RSC the existing linear axis can be used to adapt the sensor position. In cases with complex geometries other than the RSC the WLI has to be positioned and oriented by the robot during the measurement. This requires the possibility to transform the frames in an arbitrary way. The resulting algorithms for calculating the three-dimensional coordinates have to be developed to meet the criteria to keep up with the continuous measurement. In addition the influence of the robot movement has to be analyzed in respect to induced vibrations and achievable accuracy.

4.2 Time analysis

The current hardware is designed to be used as a VSI. Therefore, only a limited number of frames can be captured. After that the captured data has to be processed and only a frame with the calculated distances and one with the amplitudes are transmitted to the computer. In the new approach all frames have to be continuously passed through, because the frame stack has to be processed in a different way than with the stitching method. During transmission, no new data can currently be collected. The time t_T for transmitting the full data stack to the computer is about 1.35 s.

In order to allow for the data to be transmitted, one option would be to stop the turn table whenever the data stack is full. However, this would defeat the purpose of using a laterally scanning interferometer, which is continuous measurement. An approach that enables the interferometer to unload its data stack while maintaining an uninterrupted scanning mode, is to perform more than one 360° turn with the rotational axis. It should be noted that a time offset $t_{\text{delay}} = t_M$ is required in between each consequent turn, otherwise the same exact measurements would be performed. The measurement time for approx. 500 frames depends on the WLI parameters. Good results can be achieved with a measuring time t_M of approx. 500 ms. The time results from the frame rate and number of frames n_{Frames}:

$$t_M = \frac{n_{Frames}}{f_{Frame}} \tag{7}$$

It was decided to use the same frame rate of approx. 1000 Hz for the first tests to allow a fair comparison. As an additional assumption n_{Shift} was chosen to get a frame thickness $\Delta z'$ similar to the VSI:

$$\Delta z' = \Delta x \cdot \sin\alpha \tag{8}$$

$$= n_{Shift} \cdot s_{Pixel} \cdot \tan\alpha \tag{9}$$

The required frame thickness of approx. $\Delta\ z' = 7\,\mu\text{m}$ and (9) result in a pixel shift of $n_{Shift} = 4$ and with (2) the lateral velocity set to $v_l = 79.3\,\text{mm}\,\text{s}^{-1}$. The radius r at the corresponding ring is approx. 350 mm so the time for one ring t_{Ring} results to:

$$t_{Ring,ideal} = \frac{2\pi}{\omega} = \frac{2\pi r}{v_l} = 27.7\,\text{s} \tag{10}$$

This would be the ideal time for one ring, if the WLI was able to measure continuously. Because the time for transmitting the data is nearly three times higher than the measuring time there have to be at least four complete rotations to measure the entire ring:

$$n_{Rotations} = \frac{t_T + t_M}{t_M} = 3.7 \approx 4 \tag{11}$$

So the total time needed for one ring with the new approach could be as low as:

$$t_{Ring,LSI} = n_{Rotations} \cdot t_{Ring,ideal} = 110.8\,\text{s} = 1\,\text{min}\ 50.8\,\text{s} \tag{12}$$

In the approach with the VSI the total time t_{VSI} between two measurements is the sum of t_M and the accumulated time t_A for turning the rotation axis, waiting for the vibration to have settled and the start as well as stop ramp of the WLI itself. This results to:

$$t_{VSI} = t_M + t_A = 500\,\text{ms} + 400\,\text{ms} = 900\,\text{ms} = 0.9\,\text{s} \tag{13}$$

With a circumference of $c_{Ring} = 2\pi r = 2199.1\,\text{mm}$ and the y_{Range} from (5) the number of scans n_{Scans} per ring results to:

$$n_{Scans} = \frac{c_{Ring} \cdot (100 + Overlap)\%}{y_{Range}} \approx 416 \tag{14}$$

The total time for one ring with the VSI is:

$$t_{Ring,VSI} = t_{Scan} \cdot n_{Scans} = 374.4\,\text{s} = 6\,\text{min}\ 14.4\,\text{s} \tag{15}$$

This implies a reduction in time of 70.4 % and ideal 92.6 %.

4.3 Further challenges

The described calculations assume that the pose of the WLI can be adjusted exactly. Because the WLI is handled by a standard industrial robot the inaccuracy of the robot, as well as the calibration steps for the rotational axis and the tool center point of the WLI could lead to wrong pixel mappings while extracting the correlogram. A first approach for an automated calibration of LSIs was introduced by Munteanu [6]. These methods should be implemented and extended to meet the criteria for the use with a robot-guided WLI.
Because the LSI does not point orthogonally to the surface, shadowing could be a problem at steep slopes. Lambelet et al. has presented a new design for a WLI that could solve this issue, if measuring such geometries is required [5]: instead of a reference mirror, a skewed grating is used. Therefore, the coherence plane is rotated without positioning the WLI at an angle to the surface.

5 Conclusion

It was shown that the Olszak method could also work for RSCs. First results have been presented and the difficulties have been discussed. If it should prove to be possible to solve the described problems, the result would be enormous time savings. Additionally fewer vibrations would be induced into the system, because of the continuous measurement. The next steps should be as follows:

- Analysis of the three adaptation concepts to handle the main disadvantage of the low measurement range
- Development of fast algorithms for shifting the frames and calculating the peak of the corellogram
- Integrate an automatic calibration procedure to allow a fast change between different parameter sets

References

1. T. Domaschke, T. Schüppstuhl, and M.-A. Otto. Robot guided white light interferometry for crack inspection on airplane engine components. *ISR/Robotik, München*, 2014.
2. T. Dresel, G. Häusler, and H. Venzke. Three-dimensional sensing of rough surfaces by coherence radar. *Appl. Opt.*, 31(7):919–925, Mar 1992.
3. M. Ernst and T. Ullrich. Verfahren und Vorrichtung zur Rissprüfung eines Flugzeug- oder Gasturbinen-Bauteils, November 29 2012. DE Patent App. DE201,110,103,003.
4. G. S. Kino and S. S. C. Chim. Mirau correlation microscope. *Appl. Opt.*, 29(26):3775–3783, Sep 1990.
5. P. Lambelet and R. Moosburger. Fast and accurate line scanner based on white light interferometry. *Proc. SPIE*, 8788:87880Q–87880Q–9, 2013.
6. F. Munteanu. Self-calibrating lateral scanning white-light interferometer. *Appl. Opt.*, 49(12):2371–2375, Apr 2010.
7. A. Olszak. Lateral scanning white-light interferometer. *Appl. Opt.*, 39(22):3906–3913, Aug 2000.
8. M. Otto, T. Domaschke, S. Bahr, and M. Schnell. Präzise Rissdetektion und -bewertung an Flugzeugturbinenlinern mittels Weißlichtinterferometrie. *Automation 2014, Düsseldorf: VDI-Verl. Baden-Baden*, 2014.

Towards a Mobile Robotic Assembly System Using a Compliant Robot

Ulrich Berger, Duc Tho Le, Wenchao Zou and Mayur Andulkar

Chair of Automation Technology

Brandenburg University of Technology Cottbus-Senftenberg

Cottbus, Germany

{Ulrich.Berger, Ductho.Le, Wenchao.Zou, Mayur.Andulkar}@b-tu.de

Abstract. In this paper the use of a mobile lightweight robot is evaluated to perform an assembly task while simultaneously moving. The motion of the mobile platform results into a variable end effector position in space. The paper assesses the existing method of placing a screw in an assembly where the location of placement is variable due to end effector movement. Experiments have been conducted to evaluate the task performance by monitoring the applied force on the end effector, the position data and the task time. The results show that with impedance configuration, a moveable compliant robot is a possible solution for use in assembly operation.

Keywords: mobile robotic assembly; screwing task; compliant robot.

1 Introduction

The task of performing assembly operations on a conveyor based system is common in industries where mostly a human operator performs different assembly tasks. The application of a conventional fixed robot system (where the robot is mounted on a base or wall) to perform the assembly tasks on such a moving conveyor is impractical because of limited work space for the robot to perform the given task. Also, the movement of the components results in a dynamic environment which complicates the procedure further. Moreover, the required tolerances and the close contact between different components for assembly tasks makes the application of stiff robot system very complex. One promising solution to this problem is the use of a mobile robotic system (manipulator mounted on a mobile platform) which would reduce the non-value added process of transportation and at the same time improve the degree of automation in industries [1].

The use of a mobile robotic system for application in assembly tasks has been evaluated in our previous work. A stable synchronization between a mobile platform and a continuous moving object was developed and experiments were performed. However, a standard deviation of 2 mm was reported between the mobile platform and the object which poses challenges for the system to be applied to assembly tasks requiring low tolerance [2]. In such cases, a stiff industrial robot (without compliant configuration) would be incapable to adapt itself to the variability in positions resulting into failure or damage of assembly parts. To avoid such case, the use lightweight of compliant robots [3] mounted on a mobile platform would be a possible solution as show in Fig. 1. The compliant mode enables to continuously monitor the force and torque which is applied on the robot and reacts to resistance when work pieces do not fit properly with each other [6].

Fig. 1 Complaint robot on a mobile platform performing operation on a moveable object

This paper evaluates the feasibility to use a compliant mobile robot for performing an assembly task while the robot and the assembly platform are moving simultaneously in a synchronized manner. The synchronous motion of the robot and the assembly could save the transportation time resulting in better productivity and improve the degree of automation in industries. The motion of the robot (although on a planar surface) results in vibration of the robot base and in turn variable end effector position in space. The standard deviation in this end effector position with respect to a fixed point in space (world coordinate system) was found to be approximately of 2 mm in our previous work [2]. Moreover, the motion of the assembly results into continuously changing placement point for the object. This makes the task performance a complex procedure and poses challenges in performing the task of assembly which is assessed in this paper using the existing methods.

2 State of the art

Industrial robots have been successfully used for performing assembly operations in several cases [5] [8] [9]. Amongst the assembly tasks, the screwing task of placing a bolt in an assembly is a complex operation as it involves not just bolt placement but at the same time its rotation. This task of screwing has been studied in [4], where a robot with force sensors and multiple cameras performs screwing task. Based on the moment of force between the bolt and the nut, the position of the nut is corrected to find the best fit. In [7] a compliant reaction primitive has been built so that the unscrewing task can be performed well when the gripper rotation axis and the cap rotation axis doesn't match each other. The above work reports that the insufficient accuracy of the existing robots can be compensated with compliant configuration. However, the two systems were built for static cases.

This inspired the work on the current paper where the application of a compliant robot in a dynamic case (i.e. the robot and the assembly both are in motion) is evaluated. The challenges in performing the task of assembly are assessed for screwing operation which is taken as an example in this paper for evaluation. The screwing parameters (screwing torque limit, interrupt force and twist angle) and the existing assembly method of grasping, searching the assembly plane and performing the task are considered for experimental validation.

3 Assembly Procedure for Screwing Operation

The use of complaint robot to perform a screwing task requires consideration of several parameters. It is assumed here that the bolt is already gripped in the robot gripper and that the bolt is grasped firmly and its axis is perpendicular to the end effector of the gripper at all times during the execution of the process as shown in Fig. 3.

The procedure of performing a screwing task is described in Fig. 2. Initially the tilt of the work plane w.r.t robot base plane is found using three-point referencing. The three-point referencing is basically finding a plane and its corresponding normal using the force detection capability of the complaint robot [12]. Then the bolt is grasped from its initial position and taken to the placement position. Once the bolt (already grasped in the end effector) is brought near to the placement area as shown in Fig. 3, the gripper moves perpendicular to the work plane till the bolt touches the nut (For this purpose an interrupt force of 5N with a Custom Reference Frame (CRF) was assumed as discussed in following section). Later compliant sinking movement in the Z-axis of CRF with a Lissajou pattern on the XY-plane of CRF was executed. With impedance control, the bolt collided slightly on the boundary of the nut hole during sinking and thus fitted itself in the hole. After that, the screwing operation of the bolt around the Z-axis of CRF was performed until the bolt screwed till desired torque on the nut.

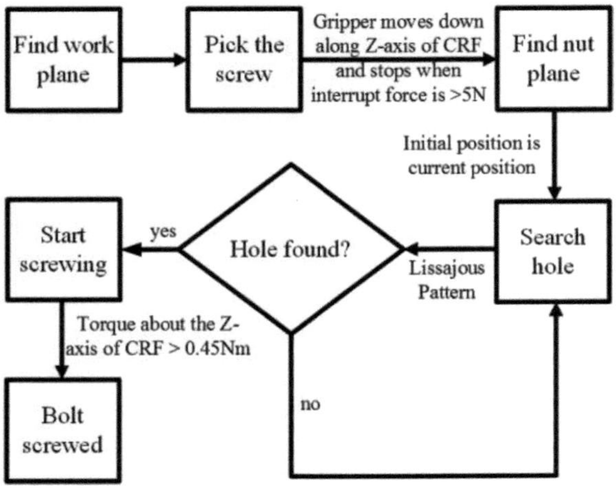

Fig. 2 Screwing task procedure

2.1 Screwing parameters

In this paper, some default parameters have been used which have been described in described in detail in [13]. These parameters can be adjusted depending on different objects.

Screwing Torque Limit: In this paper, a torque limit of 0.45 Nm around the bolt axis during screwing operation was assumed. The screwing behavior or the rotation during screwing will be terminated when this limit is reached in order to avoid damage caused by mismatching of the threads.

Interrupt Force: (Assumed as 5N). This force was defined along the Z-axis of CRF in order to detect the planes when the parts collided.

Twist Angle: (assumed as 180 degrees). The angle by which the robot will twist the bolt in each screwing movement.

2.2 Screwing steps

The steps of the screwing task are described as following:

Grasping: Initially a custom reference frame (CRF) for the work plane of the assembly was defined as the same as World Coordinate System. The Z axis of this CRF was then changed by coinciding it with the normal vector $\vec{n}$ of the work plane [10]. The gripper was used to find arbitrarily three points on the plane by using Interrupt force. Two vectors $\vec{v}_1$ and $\vec{v}_2$ were derived from the three points and further the normal vector was derived using the cross product of the vectors as seen here:

$$\vec{v}_1 \times \vec{v}_2 = \vec{n} \tag{1}$$

Once the CRF is defined, its Z axis would be perpendicular to the work plane or assembly plane and the relative position between the gripper fingertip and the work plane can be computed.

Search Nut Hole: The gripper moves perpendicular to the hole until the bolt in the gripper lightly collides with the nut plane. Consequently, the robot moves the gripper again down with a Lissajous pattern (Fig. 3) with impedance configuration to find the hole [15]. The movement of the gripper in Z-Axis of CRF decides if the hole is found.

Screw in: When the nut hole has been found, the robot rotates its end effector around Z-axis of the CRF to screw the bolt into the nut. The screw Torque limit is applied to each screwing movement. Each screw movement consists of a ±180 degrees' rotation around Z-axis with the end effector perpendicular to the work plane of the WRS. This screw movement of 180 degrees may not be possible in certain cases because of limits in robot joint angle and its corresponding configuration. This could be solved by using the null space capability of the robot [14].

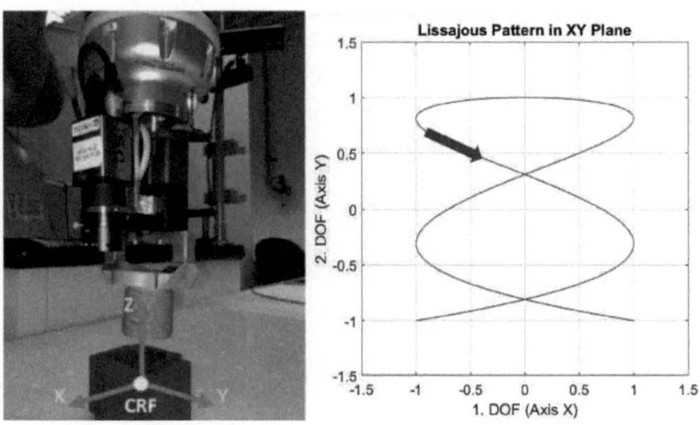

Fig. 3 The Lissajous pattern to find the nut hole

2.3 Compliant mode setting

The compliant mode is used almost for every behavior in the procedure to avoid damage of the bolt and nut, particularly for the screwing movement. In general, the bolt axis and the end effector axis may not be coincident. This would result in a shearing force on the threads during the screwing operation. The shearing force may damage the threads during this operation depending on the bolt and nut material. This shearing force f applied on the thread when they mismatch can be approximately calculated with the following equation [11], where T is the applied torque on the bolt and d_p is the pitch diameter of the bolt.

$$f = 2T / d_p \tag{2}$$

This force applies mostly on the cross section of the bolt and the nut which is parallel to the X-Y plane of the CRF. As the components where made of plastic, the damage from the shearing force was reduced by setting the robot stiffness in X-Y plane of CRF with very low value (<1 N/m) [13]. This also helped avoided the threads mismatching while screwing.

4 Experimental Setup

The experiments were conducted using a KUKA iiwa robot equipped with a two-fingered gripper was mounted on the tool flange of the robot as shown in Fig. 4. The bolt and nut were made from plastic and therefore, the parts can be damaged more easily than metal parts. The experiments were performed with the KUKA iiwa robot being fixed on a moveable platform. The variability in the experiments performed on the fixed robot was accounted by oscillating the object in the work plane with a fixed amplitude of its position. The screwing task was performed on different work planes as shown in Fig. 4.

Fig. 4. Execution of the screwing task on horizontal and arbitrarily tilted plane.

5 Evaluation Based on Results

Ten experiments each were conducted for different work plane positions w.r.t. robot base: parallel to robot base, perpendicular to robot base and two arbitrary positions of the work plane. The amplitude of oscillation was taken randomly from 2 mm to 5 mm for every experiment. This value was chosen based on the assumption that the position of the assembly placement point is known with this accuracy. This position of placement could be found using vision system in our previous work [12] or a fixed coordinate frame could be defined if possible. As seen in Table 1, the success rate in finding the hole was found to be 80%. A 100% success rate was not found because of the fact that the robot was in compliant setting and in certain configurations (e.g. near to the joint limits) the desired behavior of robot is not possible. The mean time for finding the hole was 14.81 seconds. The movement of tool center point (TCP) in X-Y plane of CRF for fixed position of nut was recorded as shown in Fig. 5.

Table 1. Experimental results for assembly operations.

Work plane orientation*	A	B	C	D
Assembly success rate (%)	80%	70%	80%	90%
Mean time for finding the hole (sec)	14.24	13.5	16.4	15.1
Amplitude for oscillation of the nut (mm)	4	3	5	2

* defined w.r.t robot base

A = parallel to robot base plane; B = perpendicular to robot base plane; C and D are arbitrarily oriented planes w.r.t. robot base

The results show that during the screwing task, the TCP moved regularly back and forth because the TCP axis and bolt center axis were non collinear due to the forced oscillations on the nut. The maximal movement of the TCP in X-Y plane of CRF amounted to 7 mm ± 1mm. This deviation was on the upper bound to the sum of the deviation from the mobile robotic assembly system (2 mm found in our previous work [2]) and the deviation in object position (2 mm to 5 mm).

It can be seen from the results that the compliant robot on a mobile platform could be used to perform a complex task of assembly even in a dynamic scenario. The deviations in TCP of the robot also direct that the use of a conventional industrial robot for such a case would be complex and would not be possible without the use of compliant motion.

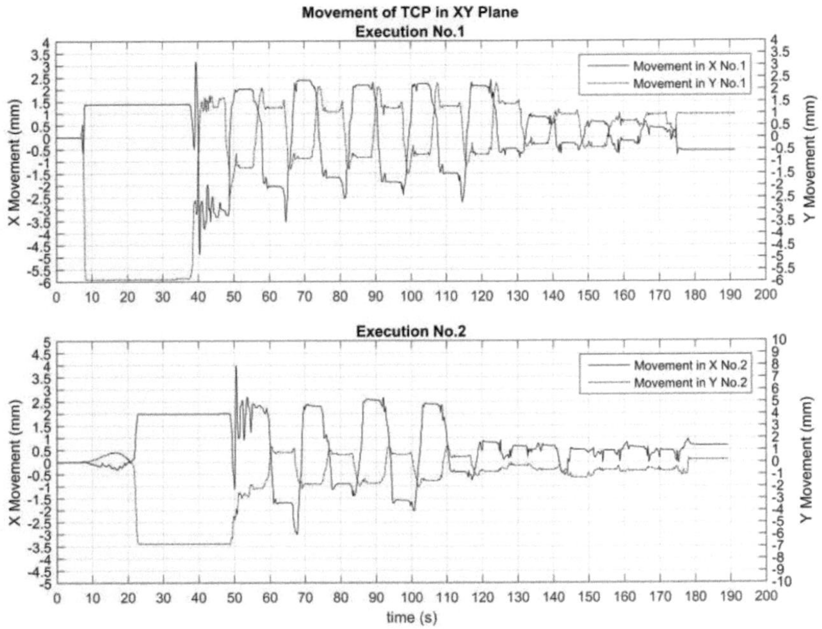

Fig. 5. Execution of the screwing task on: (top) work plane parallel and (bottom) plane arbitrarily tilted to robot base.

6 Conclusion and Future Work

In this paper, the feasibility in application of a mobile robot to perform a continuous assembly operation of screwing was evaluated. It was found that the use of conventional industrial robot for such application was not feasible because of inaccurate robot position and variability in assembly position. The use of a compliant robot for such operation was found feasible. The results indicate that with impedance configuration, a moveable compliant robot can be used to perform assembly tasks such as screwing. However, it was found that an assembly task success rate of 80% was observed which is not acceptable for industrial scenario. This could be improved by reduction in object positioning error, better robot configuration during assembly and dynamically changing the stiffness and damping parameters during assembly. The future work will focus these aspects and also on conducting experiments for other assembly operations such as fitting, gluing etc. and performing experiments in real case scenario.

Acknowledgments

The reported work was a part of the KUKA Innovation Award 2016– Flexible Manufacturing Challenge. The support of KUKA AG for providing the KUKA FlexFellow is appreciated.

References

1. U. Berger, D.T. Le, W. Zou, C. Lehmann, J. P. Städter, A. Ampatzopoulos, "Development of a Mobile Robot System for Assembly Task on Continuous Conveyor," in IN-TECH 2015 International Conference on Innovative Technologies, Dubrovnik, vol. 9, pp 11-09, 2015.
2. W. Zou, D. T. Le, U. Berger, M. Andulkar, A. Ampatzopoulos, "Integration of a PID Control System for a Mobile System for Manufacturing Task on Continuous Conveyor," in 8th IFAC Conference on Manufacturing Modelling, Management and Control, 2016.
3. A. A. Schäffer, O. Eiberger, M. Grebenstein, S. Haddadin, C. Ott, T. Wimböck, G. Hirzinger, "Soft robotics, from torque feedback-controlled lightweight robots to intrinsically compliant systems," in IEEE Robotics and Automation Magazine, vol. 15, no. 3, pp 20-30, 2008.
4. R.L.A. Shauri, K. Saiki, S. Toritani and K. Nonami, "Sensor Integration and Fusion for Autonomous Screwing Task by Dual- Manipulator Hand Robot," in Procedia Engineering, vol. 41, pp. 1412 – 1420, 2012.
5. Jörg, Stefan, Langwald, J., Stelter, J., Hirzinger, G., and Natale, C, "Flexible robot-assembly using a multi-sensory approach." IEEE International Conference on Robotics and Automation, 2000.Vol. 4, pp. 3687-3694 IEEE, 2000.
6. S. Shepherd, A. Buchstab, "KUKA Robots On-Site," in Robotic Fabrication in Architecture, Art and Design 2014, Part IV, pp 373-380, 21 March 2014.
7. J. Felip and A. Morales, "A solution for the cap unscrewing task with a dual arm sensor-based system," in Humanoid Robots (Humanoids), pp 823 - 828, Madrid, 18-20 Nov. 2014.
8. R. Bischoff, J. Kurth, G. Schreiber, R. Koeppe, A. Albu-Schäffer, A. Beyer, G. Hirzinger, "The KUKA-DLR Lightweight Robot arm-a new reference platform for robotics research and manufacturing" in 41st international symposium on Robotics (ISR) and 2010 6th German conference on robotics (ROBOTIK) , pp. 1-8. VDE, 2010
9. Wang, Lihui, Shadi Keshavarzmanesh, Hsi-Yung Feng, and Ralph O. Buchal. "Assembly process planning and its future in collaborative manufacturing: a review." The International Journal of Advanced Manufacturing Technology 41, no. 1-2 (2009): 132-144.
10. A. J. Hanson, "Geometry for n-dimensional graphics," in Graphics Gems IV, pp. 149-170, 1994.
11. P. R. N. Childs, "Mechanical design engineering handbook." Butterworth-Heinemann, 2013.
12. D. T. Le, M. Andulkar, W. Zou, J. P. Staedter, and U. Berger, "Self Adaptive System for Flexible Robot Assembly Operation," in 2016 IEEE International Conference on Emerging Technologies and Factory Automation (ETFA), 2016.
13. Lebedynska, Y.: "Entwicklung eines Informationssystems mit Reifegradmanagement für automatisierte Produktionsprozesse", Diss., ISBN-Nr. 978-3-8440-0289-8 ,Shaker Verlag Aachen, 2011

14. Lehmann, C.: Entwicklung eines fehlertoleranten interaktiven Programmier- und Bedienkonzeptes für wandlungsfähige Industrieroboterzellen, Diss., ISBN-Nr. 978-3-8440-4096-8, http://dx.doi.org/10.2370/9783844040968, Shaker Verlag Aachen, 2015.
15. Jokesch, Michael, Suchý, J., Winkler, A., Fross, A., & Thomas, U.: "Generic Algorithm for Peg-In-Hole Assembly Tasks for Pin Alignments with Impedance Controlled Robots." Robot 2015: Second Iberian Robotics Conference. Springer International Publishing, 2016.

Task-based Simulation Tool for Human-Robot Collaboration within Assembly Systems

Paul Glogowski[1], Kai Lemmerz[1], Lena Schulte[2], Andre Barthelmey[2], Alfred Hypki[1], Bernd Kuhlenkötter[1], and Jochen Deuse[2]

[1] Ruhr-University of Bochum, Chair of Production Systems
Universitätsstr. 150, 44801 Bochum, Germany
glogowski@lps.rub.de
WWW home page: http://www.lps.ruhr-uni-bochum.de/
[2] TU Dortmund University, Institute of Production Systems
Leonhard-Euler-Str. 5, 44227 Dortmund, Germany

Abstract. The human-robot collaboration unites positive skills of humans (flexibility, intuition, creativity) and robots (strength, stamina, velocity, precision) in order to ensure a high level of resource efficiency and productivity also with smaller batch sizes and a high variety of versions in production. However, the design of a collaborative assembly system is a complex engineering challenge due to the different goal criteria that have to be considered. A simulation tool for designing and securing human-robot collaboration is yet missing which prevents the widespread use of such technologies. This paper proposes a concept of a simulation tool for collaborative workplaces. In the future, this will ensure that also companies with little experience and limited resources are empowered to implement human-robot collaboration systems prosperously.

Keywords: Human-Robot Collaboration, Collaborative Assembly Systems, Simulation Tool, Distribution of Tasks on Humans and Robots

1 Introduction

Today's industrial assembly is facing specific demands due to individual and complex products, fluctuating quantities, ever decreasing product life cycles as well as new process requirements. Since both manual and fully automated assembly systems do not meet the heterogeneous requirements in its entirety, it has become necessary to develop, design and implement flexible systems for assembling complex products. Human-robot collaboration (HRC) and secure layouts of workplaces in assembly are central in this context. [1]

The design of a manual assembly workplace without a collaborative system is already a complex engineering challenge in terms of work content and working hours. Due to the variety of necessary goal criteria such as automation, technical and economic suitability, ergonomics and safety, this process is nowadays usually realised by involving a specially developed software for the modelling, simulation and optimisation of the workplaces. In use of workplaces, in which humans and robots collaborate, there have not yet been any simulation tools which could simulate or evaluate these systems, taking the above points into account. Thus, the required modelling effort to determine the potential of collaborative assembly systems, to evaluate the technical-economic feasibility and to plan and design the systems is immense and poses an essential impediment for using human-robotic systems.

This paper proposes a concept of an overall simulation tool for collaborative assembly systems in combination with multiple product scenarios. In the future, this will ensure that also companies with little experience and limited resources are empowered to implement collaborative systems successfully even if the diversity of products and variants fluctuates. Process planners will be supported in the planning and designing of a HRC scenario. Furthermore, an evaluation feature will be integrated into the simulation tool so that during the future planning of an assembly scenario, assembly tasks can be optimally distributed as so-called subtasks to the capabilities of humans and robots.

The paper is organised as follows: Section 2 describes a short state of the art in the area of HRC and corresponding frameworks for the modelling and simulation of industrial assembly applications. Section 3 and 4 present an approach of a task-based simulation tool for collaborative assembly systems. Thereby, an overall architecture of the simulation tool as well as its subcomponents are stated in detail. Finally, we conclude with a summary and provide an outlook for future research dealing with a HRC in section 5.

2 State of the Art

2.1 Human-Robot Collaboration

The use of HRC makes it possible to combine the learning ability and flexibility of humans with the advantages of modern robots. The main focus is on increasing productivity, improving the quality of work as well as reducing the workload for employees. Especially in assembly, the robot functions primarily as the employee's assistant, i. e. the robot undertakes subtasks and/or repetitive processes at the workplace that are inconvenient for humans in terms of ergonomics. On the other hand, the human employees contribute to the HRC through their experience, fine-motor skills as well as attributes such as creativity, flexibility and power of judgement. Thus, in collaborative assembly workplaces, human employees will still play a decisive role in the future [2].

Legal conditions for the admission of direct HRC were established through changes and new developments concerning official standards and directives. These include e. g. the norms ISO 10218 [3, 4], the technical specification ISO/TS 15066 "Robots and robotic devices – collaborative robots" [5] published in February 2016 and the BG/BGIA risk assessment recommendations regarding the design of workplaces with collaborative robots [6]. These normative conditions are necessary in order to guarantee the safety of the human employees in a collaborative environment. It is possible to deduce appropriate security strategies from these norms and directives. In this context, it is essential to assess relevant hazard potentials and to determine the costs and consequences of possible safety precautions in planning applications with HRC.

Despite a high economic potential and the resulting improvement in ergonomics, collaborative assembly systems have not yet become widely prevalent due to aspects of safety and product liability as well as insecurities concerning system integrators. However, initial research and industrial projects show approaches for the integration of HRC [7, 8]. Furthermore, numerous systems (e. g. ABB YuMi [9], KUKA LBR iiwa [10], Bosch APAS family [11]) have been developed according to current safety requirements of HRC.

2.2 Design and Simulation of Human-Robot Collaboration in Assembly

For a broad use of a HRC in assembly, a simulation tool is required which allows the easily accessible planning and securing of HRC scenarios. Currently, there are various efficient simulation tools which are either exclusively specialised in human simulations or robot simulations. However, none of them reproduce HRC systems entirely.

Human simulation systems. There are specific simulation tools available which focus on different planning tasks and display mostly anthropometric human models in manual work systems [12]. The *Editor for Manual Work Activities* (*ema*) of the company imk automotive GmbH covers both a temporal and ergonomic evaluation of the planned workflow as well as a mostly realistic visualisation of the human body movement [13, 14]. However, it has not yet been possible to integrate a sufficient functionality for simulating components of automation technology, especially in collaboration with the human model.

Robotic simulation systems. Robot simulation systems, especially those used for offline programming, can be divided into manufacturer-specific or universally usable systems. Universally applicable systems generally utilise robotic libraries with components of multiple manufacturers. [15] Apart from a virtual image of the work system, the robots' motion behaviour can be displayed within the simulation's framework and, among others, be analysed regarding potential collisions and execution time. Human simulations or even HRC simulations are, however, not entirely supported within these simulation systems.

HRC simulation systems. The manufacturer-independently usable software *FAMOS robotic* offers the prototypical integration of a digital human model for developing HRC as a welding task assistance [8, 16, 17]. Furthermore, a planning and a simulation system for transferring manual processes to automatic or semi-automatic solutions of service robotics were developed in previous research work [18, 19]. However, it has not yet been possible to provide a digital tool which would realise a comprehensive simulation of a HRC in the whole area of the assembly process.

3 System Architecture

The following chapter describes the simulation tool for HRC, so that both humans and robots can operate side by side in a collaborative assembly process. Figure 1 shows the overall architecture of the simulation tool for collaborative assembly systems.

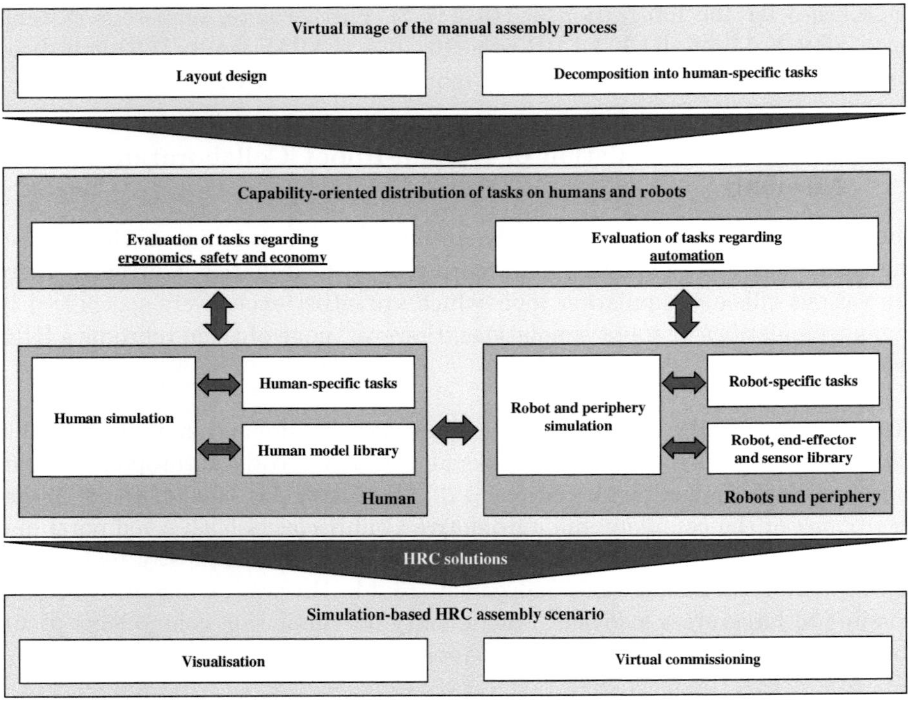

Fig. 1. Scheme of the task-based simulation tool

In this paper, a manual assembly process in the form of a virtual image represents the initial situation for the simulation tool. Thereby, the virtual image includes all necessary input data for the simulation tool such as assembly schedules, parts lists and geometric data of the product and layout. The virtually represented assembly scenario is examined from two different perspectives. On the one hand, the evaluation is based on human factors and considers technologies such as ergonomics, safety and economy. On the other hand, another evaluation scheme examines the assembly process according to automation engineering such as accuracy, reachability and manipulability of the possible automation components like the collaborative robot.

In terms of the manual assembly process, human body movements (e. g. standing up, running) or the handling of tools (e. g. using a tool, moving a tool) are typical human-related subtasks. The aim of the simultaneous evaluation features is to identify and verify automatable subtasks. For this purpose, new subtasks have to be created that can be processed by the automation components. The automatable subtasks can be considered on different hierarchical levels since a composition of individual subtasks is possible. An exemplary task scenario for the robot is a pick-and-place application. The robot-based task *pick-and-place* includes multiple subtasks (see Fig. 2): move to position A, close the gripper, pick the object up, move to position B, put the object down, open the gripper. During instantiation, in other words, when a task is assigned to a robot, certain basic parameters (position A and B, object selection, gripper type etc.) will be stated for this task. The single subtasks (e. g. move to position A) access these stored basic parameters.

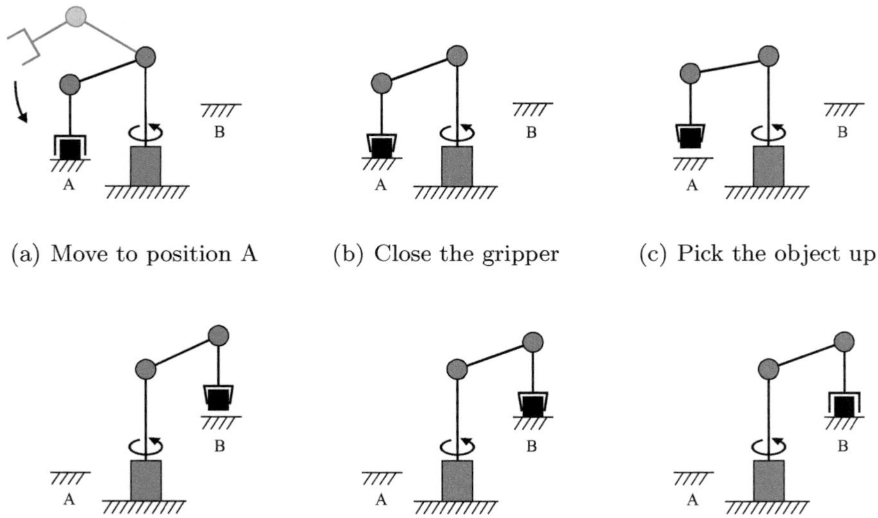

(a) Move to position A (b) Close the gripper (c) Pick the object up

(d) Move to position B (e) Put the object down (f) Open the gripper

Fig. 2. Robot-based task *pick-and-place*

In order to identify automatable subtasks, critical human-related subtasks are examined whether and how they can be supported by automation components so that a direct collaboration between humans and robots takes place. For example, the robot can physically relieve the employee by assisting in ergonomically inconvenient handling of workpieces.

Related to the different goal criteria such as automation, technical and economic suitability, ergonomics and safety, the resulting procedure provides suggestions of HRC solutions, in the form of an assignment of specific subtasks to automation components. The execution of these subtasks will be simulated by means of possible automation components from a robot and component library and checked for technical feasibility. This enables assembly scenarios with different degrees of automation and different automation components, depending on the applied robots and the distribution of assembly subtasks on humans and robots. The simulation tool is used to secure and evaluate these assembly scenarios in an integrated simulation environment. As part of the integral simulation, the HRC will be visualised and thus can be secured prospectively concerning collisions and other hazards for human employees. Subsequently, the process planner can select the most suitable assembly scenario according to the different goal criteria.

4 System Components

In the following chapter, the main components of the simulation tool for a HRC are described in detail. The system components are the following (see Fig. 1): the robot and component library, a corresponding simulation environment for HRC-capable robots and peripheral components, a human simulation system as well as an interdisciplinary evaluation scheme for the analysis and assessment of the HRC workplace with regard to aspects of ergonomics and automation.

Robot, end-effector and sensor library. One component of the simulation tool is the robot, end-effector and sensor library for collaborative assembly solutions which is independent from any manufacturer. The physical system properties or parameters of the various robots and peripheral objects are stored in a database in order to build up a component library. The kinematic structure (Denavit-Hartenberg parameter, joint topology etc.), the geometry (masses, inertia, center of gravity etc.), performance characteristics (payload, engine power of the axes, end-effector velocity and acceleration etc.) or structural properties (stiffness and damping values, friction coefficient etc.) of a robotic system are an exemplary set of parameters.

Offline-based simulation of robots and periphery. The main focus of the simulation tool is on the simulation environment for robots and peripheral components. The robot and periphery simulation generates important input data for the evaluation scheme and for the capability-oriented distribution of subtasks on humans and robots. In this context, the simulation handles the robot-specific processing data (end-effector velocity and acceleration, process forces etc.), generates the resulting robot and end-effector motion and provides it for the cooperating human simulation. A physical model of the motion and structural behaviour (inverse kinematics, forward kinematics, dynamics etc.) of robot and peripheral components is derivated and suitably assigned to a visualisation environment, e. g. through the representation of the robot's path and workspace. Within the robot and periphery simulation, the selected components are checked and examined for their technical feasibility with respect to their system properties. Therefore, the simulation uses the object properties stored in the robot and component library. For a safe design of the HRC workplace, the forces and moments that result from the robot and peripheral components and which act on the human employee will be simulated. In this way, the force limits defined in the technical specification ISO/TS 15066 can already be considered in the simulation. In addition, a collision check and monitoring function inside the simulation environment is part of the simulation tool which makes a simulation-based HRC possible.

Offline-based simulation of humans. Another subsystem deals with the area of human simulation, but is covered by existing simulation tools, such as the *ema* software tool. The interaction of the simulation tool with the human simulation is carried out via a data interface, which is connected to the evaluation scheme and to the robot and periphery simulation. The human simulation provides all the human movement data (e. g. joint data at any time) for a specific subtask. From this, the movement of the human workers will be simulated. In order to avoid collisions between robots and humans, the human simulation results will be used for path planning of the robots.

Capability-oriented distribution of tasks on humans and robots. This subcomponent of the simulation tool is used for the capability-oriented classification into manual and robot-based assembly subtasks, including a valuation of ergonomics and automation technology. The logic for the capability-oriented distribution of subtasks uses the components of the robot, end-effector and sensor library in order to provide a suitable selection of automation components for a particular assembly task. These results serve to review and evaluate the selected automation components. Afterwards, each subtask in the assembly process can be optimally distributed to the capabilities of humans and robots. In addition, the capability-oriented classification of the assembly task includes a derivation of the required support for the human assistant. Depending on the skills and qualifications of the assistant, the capability-oriented distribution provides a suitable robot.

5 Summary

The paper at hand introduces a concept of a task-based simulation tool for collaborative assembly systems. Thus, it contributes especially to the empowerment of companies with little experience and limited resources regarding the use of HRC systems.

The overall architecture of the suggested simulation tool consists of two main functional parts: a human simulation as well as a robot and periphery simulation. The simulation of subprocesses carried out by human employees is provided by the *Editor for Manual Work Activities* (*ema*) of the company imk automotive GmbH, which offers the necessary human simulation functionality. For simulating the automated steps in HRC scenarios, a manufacturer-independent library of automation components (robots and periphery) will be developed and applied for this area of simulation. By coupling these two simulation components, means designing a concept for data formats as well as for data exchange and consistency, the subtask allocation for different HRC solutions can be simulated and evaluated.

After the successful simulation of collaborative assembly systems, the simulation results and the visualisation of the overall process including all human and automated parts are an ideal base to develop a suitable concept for the qualification of the workers, e. g. based on drawings, screenshots or higher level usability videos. The concept to be developed within the current research project "KoMPI" will support design processes as well as the ongoing operation of HRC systems. In this context, it is essential to consider the design of the human-machine interface as well as specific trainings to guarantee qualified staff.

Acknowledgment

The research and development project "KoMPI" is funded by the German Federal Ministry of Education and Research (BMBF) within the Framework Concept "Research for Tomorrow's Production" (fund number 02P15A060) and managed by the Project Management Agency Forschungszentrum Karlsruhe, Production and Manufacturing Technologies Division (PTKA-PFT).

References

1. Lotter, B.; Wiendahl, H.: Montage in der industriellen Produktion. Berlin Heidelberg: Springer Vieweg (2012).
2. Schröder, N.: Robotik-Guide 2016. In: Mechatronik 2016, No. 6, pp. 36-38 (2016).
3. ISO 10218-1:2011 Robots and Robotic Devices - Safety Requirements for Industrial Robots - Part 1: Robots.
4. ISO 10218-2:2011 Robots and Robotic Devices - Safety Requirements for Industrial Robots - Part 2: Robot Systems and Integration.
5. ISO/TS 15066:2016 Robots and Robotic Devices - Collaborative Robots.
6. BG/BGIA Risk Assessment Recommendations according to Machinery Directive: Design of Workplaces with Collaborative Robots (2011).
7. Zwicker, C.; Reinhart, G.: Human-Robot-Collaboration System for a Universal Packaging Cell for Heavy Electronic Consumer Goods. In: Enabling Manufacturing Competitiveness and Economic Sustainability , pp. 195-199 (2013).
8. Thomas, C.; Matthias, B.; Kuhlenkötter, B.: Human-Robot-Collaboration - New Applications in Industrial Robotics. In: International Conference on Competitive Manufacturing (COMA), Stellenbosch, pp. 293-299 (2016).
9. ABB YuMi. http://new.abb.com/products/robotics/de/industrieroboter/yumi.
10. KUKA LBR iiwa. https://www.kuka.com/en-de/products/robot-systems/industrial-robots/lbr-iiwa.
11. Bosch APAS. http://www.bosch-apas.com/de/apas/start/bosch_apas.html.
12. Schönherr, R.: Simulationsbasierte Absicherung der Ergonomie mit Hilfe digital beschriebener menschlicher Bewegungen. p. 133 (2013).
13. Fritzsche, L.; Jendrusch, R.; Leidholdt, W.; Bauer, S.; Jäckel, T.; Pirger, A.: Introducing ema (Editor for Manual Work Activities) – A New Tool for Enhancing Accuracy and Efficiency of Human Simulations in Digital Production Planning. In: Digital Human Modeling, HCII, Berlin, Heidelberg, Springer-Verlag, pp. 272-281 (2011).

14. Bauer, S.: Prozesssprachenbasiertes System zur Ansteuerung digitaler Menschmodelle als Teilkomponente einer Software zur Planung und Visualisierung menschlicher Arbeit in der Digitalen Fabrik. (2015).
15. Hypki, A.: Beitrag zur Simulation industrieller Automatisierungssysteme. Shaker Verlag GmbH (2009).
16. Busch, F.: Ein Konzept zur Abbildung des Menschen in der Offline-Programmierung und Simulation von Mensch-Roboter-Kollaborationen. Shaker Verlag GmbH (2016).
17. Busch, F.; Thomas, C.; Deuse, J.; Kuhlenkötter, B.: A Hybrid Human-Robot Assistance System for Welding Operations - Methods to Ensure Process Quality and Forecast Ergonomic Conditions. In: Technologies and Systems for Assembly Quality, Productivity and Customization, Proceedings of 4th CIRP Conference on Assembly Technologies and Systems "CATS", Ann Arbor (2012).
18. Klöckner, M.; Weisner, K.; Kuhlenkötter, B.; Deuse, J.; Roßmann, J.: Entwicklung eines Werkzeuges zur anwendungsfallorientierten Auswahl servicerobotischer Systeme. In: Tagungsband der VDI-Mechatroniktagung 2015, pp. 1-6, Dortmund (2015).
19. Deuse, J.; Rossmann, J.; Kuhlenkötter, B.; Hengstebeck, A.; Stern, O.; Klöckner, M.: A Methodology for the Planning and Implementation of Service Robotics in Industrial Work Processes. In: Proceedings of the 5th CIRP Conference on Assembly Systems and Technologies "CATS", 23, pp. 41-46 (2014).
20. Beumelburg, K.: Fähigkeitsorientierte Montageablaufplanung in der direkten Mensch-Roboter-Kooperation. Jost-Jetter Verlag (2005).

Centralized vs. decentralized storage of semantic definitions used for data exchange within the CAx toolchain

Matthias Bartelt and Bernd Kuhlenkötter

Ruhr-Universität Bochum, 44801 Bochum, Germany,
{bartelt, kuhlenkoetter}@lps.rub.de

Abstract. The planning of production systems involves many software systems of different kinds. Although these systems are part of different domains, there is an intersection in which the same set of information is used among several systems. Indeed, at some point during the planning, it is necessary to exchange information between the systems. Although a holistic and lossless exchange of all data between all systems may not be required, there is a need for a high interoperability of all involved systems. A main objective within this context is to provide the required semantics to describe the information to be exchanged. Since the number and kind of involved systems during planning processes is not determined, the different kinds of information are not determined either. Hence, the definitions of the corresponding semantics should be changed dynamically. Within this paper we present two methods that pursue an infrastructure with a high interoperability of arbitrary systems. Both methods use AutomationML as an intermediate format. One method uses a central storage to store semantic definitions, the other decentrally stores semantic definitions along with the corresponding system. We have realized both approaches and have defined a use case for verification. In a final comparison of both methods, we provide an outline of the major advantages and disadvantages. Even though there is still need for further investigations, a combination of both options will most likely enable a holistic data exchange within the CAx toolchain.

Keywords: Concurrent engineering, Holistic exchange of information, CAx toolchain, Interoperability, Heterogeneous system topography

1 Introduction

Once a product is designed, it is necessary to plan an appropriate production plant. This planning phase is nowadays characterized by a high level of digitization. In addition, software systems of different domains are involved within this phase. By the use of these systems, a huge amount of information of different domains is created and should be shared among the other involved systems. For this interoperability, it is essential to provide a loss-free and holistic exchange of data.

Since any software system generally uses its own database for storing information, there is need for the homogenization of data. A first approach is the so-called computer integrated manufacturing (CIM), with the concept to combine the individual databases into one [1]. However, it has not yet been possible to achieve an economical implementation of a holistic digital workflow. Reasons are given in [2] and are, amongst others, a high heterogeneity within software systems as well as high costs. Although the heterogeneity can be reduced (e.g. by using software from the same vendor), an opposing trend can be observed, which is caused, amongst others, by an increasing outsourcing practice [3].

Nevertheless, digital manufacturing has the capability to shorten development time and costs and enable decentralized manufacturing as well as a comprehensive collaboration of companies and suppliers [4]. Especially in the latter one, small and medium-sized enterprises (SME) are involved, which usually lack an appropriate implementation of a digital manufacturing process [5]. Furthermore, the software systems of those enterprises are usually part of different domains and different vendors, which is why it is not easy to guarantee a holistic exchange of data.

2 Requirements

Existing product lifecycle management (PLM) systems indeed allow the usage of the same data among different domains. Nevertheless, the planning and simulation of production systems is often done by SMEs. Additionally, it is most likely that several SMEs are involved within the engineering process. Thus, software systems of different vendors are involved which are not compatible among one another in terms of data exchange.

In order to achieve a holistic data exchange, it must be possible to interconnect all systems involved. However, even if lifecycle concepts are integrated into a software system, it still makes use of different approaches because one holistic approach is still missing [6]. Interconnections must be implemented by the SMEs themselves, since software vendors do not provide an appropriate solution to exchange the required data. This knowledge has been gained within the so-called *conexing* research project [7] which has analyzed different workflows in the context of virtual commissioning. Furthermore, many SMEs do not have the required knowledge or time for such an implementation. For this reason, a method is required which allows for a simple exchange of data with a very low implementation effort.

In search of an infrastructure for a holistic data exchange, it is essential to avoid any change of software systems if either the number and kind of involved systems or the kind of data varies. This is also a main requirement as soon as systems in operation are involved within the exchange. Since such systems like cyber-physical systems [8] or cyber-physical production systems [9] are part of an ongoing production, it is most likely that they are not allowed to be changed. Responding to this constraint, we want to facilitate a lossless, multi-directional and cross-company handling and usage of information of arbitrary kind within the engineering process. This will serve to improve the harmonization and interoperability of different software systems, which is, according to [10], still regarded as a main obstacle.

A main feature of the required infrastructure is the possibility to exchange data between different software systems without the need of a software engineer, who has a deeper knowledge of a particular system. As a matter of fact, such a software engineer is usually not available during engineering processes, which is why this feature is a main requirement of concurrent engineering in real-life planning of production systems. Especially SMEs should be able to participate in cross-company digital manufacturing processes due to the possibility of a comprehensive data exchange.

3 Using Design Pattern

At a lower level of information exchange, single data objects have to be exchanged among the involved systems. The combination of these objects defines a system-independent data set, which has to be shared, including the semantics. The Virtual Factory Framework [11], which is based on tools of the Semantic Web [12], is one example of such an approach. Nevertheless, whenever information is exchanged, the principal problem is always the same, independently of the approach: the information of one system has to be transferred to another system. This implies the fact that there are two variants of the semantics.

The semantics can be defined as the dedicated specification. They are known and the exchange of information is reduced to an exchange of data between the systems. Obviously, only the specified data can be exchanged. Thus, this approach is not suitable for a holistic data exchange. Alternatively, the data as well as their semantics are stored within an intermediate exchange format. A major issue with this approach is the definition of the semantics because different systems may store the same information while describing their semantics in a different way. However, the next chapter presents that a holistic exchange of information is indeed possible.

Having a deeper look at the data exchange between two systems, we can identify different patterns. This method to identify a generic solution for recurring problems was introduced in [13], where so-called design patterns are described: 'Each pattern describes a problem which occurs over and over again in our environment, and then describes the core of the solution to that problem, in such a way that you can use this solution a million times over, without ever doing it the same way twice.' In software engineering, design patterns are well-established, and nowadays many different kind of patterns exist (cf. [14]).

We have identified two design patterns (see fig. 1), which are suitable for the desired exchange method. As depicted in fig. 1(a), two systems S_i and S_j may store their information in a proprietary data format A_i and A_j. The exchange itself is performed by the use of an intermediate format B. If more systems are involved, they also use the same format to exchange their information.

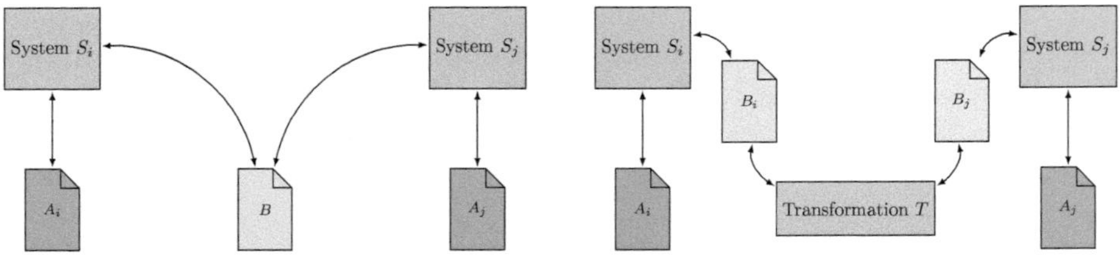

(a) Exchange with intermediate format (b) Exchange using a transformation

Fig. 1. Design pattern for exchange of information between two systems.

Another possibility is a transformation of the system's information into different intermediate formats B_i and B_j and to use a transformation T to map the information from one intermediate format to the other, as it is shown in fig. 1(b). If more systems are involved, the formats B_i and B_j as well as the transformation T are not affected, since they are dependent on the systems S_i and S_j only. Instead, new transformations must be created in order to achieve a proper mapping between the systems' information.

4 Centralized and Decentralized Storage

As stated above, we do not use exchange formats with fixed specifications. Instead, we want to realize an exchange of information instead of data. Thus, we have to store the semantics of the data also into the intermediate exchange format. Nevertheless, any system must be able to process the exchanged information. We have identified two different approaches.

The first approach facilitates a unified definition which is shared among all systems (cf. fig. 1(a)). Hence we call this option *centralized storage.* As soon as new kinds of information must be processed, this definition can be dynamically extended. Of course, the intermediate formats must be capable to describe the exchanged information semantically. As mentioned earlier, the number and kinds of involved systems are not fixed. Thus, the kind of information which must be transferred will most likely change as soon as new systems are involved. In fact, this yields to a change of the intermediate format B (i.e. its semantics) and also to a change of the embedded unified semantic definitions.

In the second approach, each system has its own intermediate format (cf. fig. 1(b)). As a main benefit of this approach, the format must only be changed if the corresponding system is changed (e.g. new software version). Also, the format can be developed independently. Nevertheless, all intermediate formats should use the same syntax in order to be able to define appropriate transformations. Since the intermediate format belongs to the corresponding system, we call this option *decentralized storage.*

4.1 Implementation

In order to compare the centralized and the decentralized approach, we have developed an implementation for a range of different software systems. The intermediate formats are based on the recently emerged AutomationML format [15], which has been developed in order to store engineering information from different domains. It combines several standards, like CAEX, PLCopen or COLLADA, and is able to describe semantic and topological information as well as logical behavior, geometry and kinematic models. All included formats are based on the XML format, which is pervasive nowadays [16]. AutomationML is object oriented and can store relationships (physical or logical) between objects (cf. [17]). A major benefit of this format is the possibility to define the semantics within the format. Thus, for any kind of object, its data and semantics can be described using AutomationML. With this, the format has the required flexibility to store any information to be exchanged.

Within the *conexing* research project, we have built an online server-based infrastructure. The server provides a SOAP interface, which is used to receive required meta-data for the exchanged information. The AutomationML documents themselves are transferred via HTTP and are managed by the server. Software systems can connect to the server and receive the current version of the desired AutomationML document. The system can import data and exports its own data to the document, while change and access management is handled by the server.

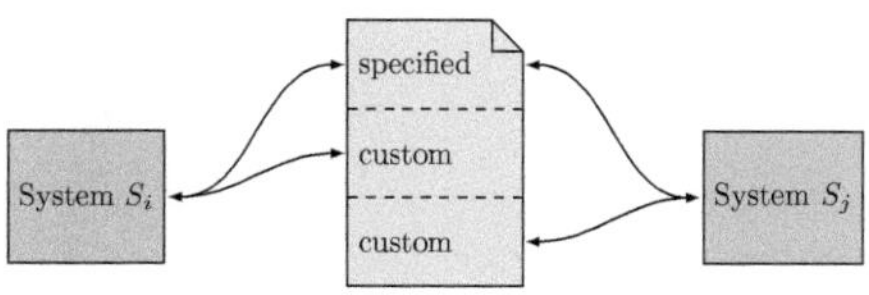

Fig. 2. Exchange of information between two systems. Only information with specified semantics can be used by both systems.

A drawback of this approach is visualized in fig. 2. Along with the information, the semantics of the corresponding AutomationML object have to be assigned. For some kind of information, e.g. geometrical information, there is a fixed specification included in the AutomationML standard. We have created further specifications to describe e.g. technical drawings. Of course, a system may export any information including semantic information. Nevertheless, other systems are only able to use information which comply with the existing specifications. As a benefit, only a low effort is required to extend a system in order to work within the created infrastructure. Moreover, we provide a software library which includes several common functionalities. With this, the effort can be reduced even further.

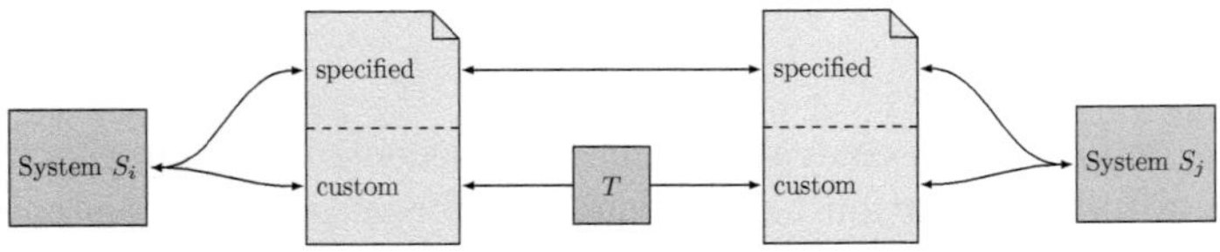

Fig. 3. Exchange of information by using transformations. Adhering to specifications will avoid some transformations.

The decentralized approach is drafted in fig. 3. We do not need a server, since each software system is responsible for its own data. Furthermore, using specified semantics may simplify the exchange, but is not required. Since we use AutomationML as an intermediate format, we can use XSLT stylesheets to transform an AutomationML document from one system in such way that another system can use the provided information. Basically, the stylesheet describes a mapping between the semantics of two software systems. Of course, a mechanism is required to merge the system's document with the transformed document. Again we provide a software library with appropriate functionalities.

4.2 Use Case

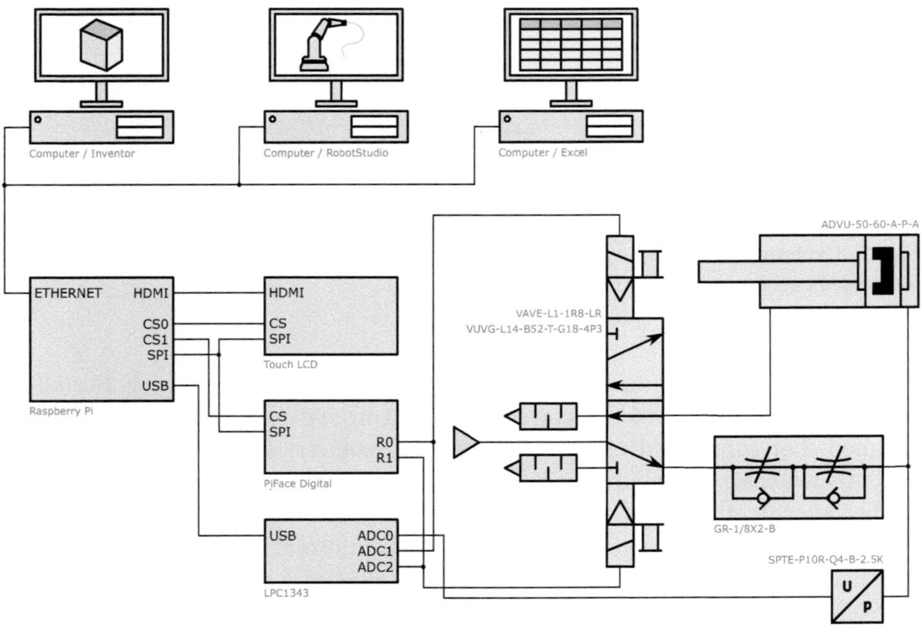

Fig. 4. Use case to verify both approaches. Three software systems of different domains as well as a cyber-physical system are connected via Ethernet.

To verify both approaches, we have connected four systems of different domains: ABB RobotStudio is used for the offline programming and simulation of industrial robots. Autodesk Inventor is used to design parts and assemblies. Since Microsoft Excel is used for different purposes, we have also created an integration of this system. Furthermore, we have developed a cyber-physical pneumatic cylinder which is also connected to the infrastructure. The resulting infrastructure is outlined in fig. 4. The systems transfer their information via Ethernet while using one of the described methods.

With these systems we have defined an industrial robot-based production cell along with a workflow for the planning of this cell. As a first step, the information of the cyber-physical cylinder is used to create a die cutter tool with Inventor. Afterwards, the resulting tool is transferred to RobotStudio and the machining task is planned. After both steps, Excel is used to create a part list of the current state of the planned system.

4.3 Comparison

With both described approaches, a holistic and lossless exchange of information can be achieved. Nevertheless, there are two major differences, as described below.

If the central storage method is applied, the same semantic definitions are shared among all involved systems. Because of this, a unified AutomationML document can be created which stores the information of all systems. In fact, if the server is used to exchange the information, the server always contains the current development state of the planned project. In case of the other method, each system only contains its own information. If the mentioned transformations are applied, a lossless usage of any information in any other system is possible but usually there is no unified document which contains the complete planning project. However, if some information is not required by a system, there is no need to create a transformation.

The second major difference is the required effort to include new systems. In both cases an initial effort is required to extend a system in order to connect to the provided infrastructure. In the centralized case, each system can directly use the provided information of the new system as long as the semantics of the information are defined. If some semantics are missing, an appropriate definition must be created and all systems which want to use this new kind of information must be updated. In contrast, the decentralized approach requires the creation of various new transformations but there is no need to change any system involved. This decentralized approach is more promising for our use case. The main reason for this is the ease of implementation of the intermediate format. The only requirement is the usage of AutomationML (in our case). No coordination with any other manufacturer is required. Additionally, the required transformations can be defined easily and individually If this approach is used, it will be rather simple and cost-efficient to provide an exchange of information between two systems.

5 Conclusion

Within this paper we have presented two different methods to exchange information among an arbitrary set of different systems. We have created an implementation for each approach and defined a use case to verify the methods. Even though this applied example describes a rather simple production cell, it involves systems of different domains. For this reason, we are sure that the methods can easily be applied to larger projects.

Up to now, we have not yet performed any economic analysis. Such an analysis will serve as a metric for an objective comparison of the presented approaches. To guarantee a broad application, it will be essential to identify the economic usage as it is a key performance indicator.

Although we have focused on software systems, it is also possible to integrate other systems, like cyber-physical systems. A holistic and lossless exchange of information can be achieved in both cases. Additionally, systems are not subjected to a software change whenever a new kind of system or new kind of information is available. However, both methods have some drawbacks, which prevent a broad usage. In future work, we will try to combine both methods in order to eliminate the disadvantages. For this purpose, we will develop an automatism to create the mentioned transformations. It should be possible to share each created transformation among the systems in order to fully or partially reuse them. If these aspects are thoroughly considered, a holistic data exchange can successfully be realized within the CAx toolchain.

References

1. Scheer, A.W.: CIM: Computer integrated manufacturing; towards the factory of the future, 3 edn. Springer, Berlin (1994)
2. Abramovici, M., Schulte, S.: PLM - Neue Bezeichnung für alte CIM-Ansätze oder Weiterentwicklung von PDM? Konstruktion - Zeitschrift für Produktentwicklung und Ingenieur-Werkstoffe (1/2) (2005)
3. Kenyon, G.N., Meixell, M.J., Westfall, P.H.: Production outsourcing and operational performance: An empirical study using secondary data. International Journal of Production Economics **171**, 336–349 (2016). DOI 10.1016/j.ijpe.2015.09.017
4. Chryssolouris, G., Mavrikios, D., Papakostas, N., Mourtzis, D., Michalos, G., Georgoulias, K.: Digital manufacturing: History, perspectives, and outlook. Proceedings of the Institution of Mechanical Engineers, Part B: Journal of Engineering Manufacture **223**(5), 451–462 (2009). DOI 10.1243/09544054JEM1241
5. Bracht, U., Masurat, T.: The Digital Factory between vision and reality. Computers in Industry **56**(4), 325–333 (2005). DOI 10.1016/j.compind.2005.01.008
6. Umeda, Y., Takata, S., Kimura, F., Tomiyama, T., Sutherland, J.W., Kara, S., Herrmann, C., Duflou, J.R.: Toward integrated product and process life cycle planning—An environmental perspective. CIRP Annals - Manufacturing Technology **61**(2), 681–702 (2012). DOI 10.1016/j.cirp.2012.05.004
7. Bartelt, M., Kuhlenkötter, B. (eds.): conexing Abschlussbericht: Werkzeug zur interdisziplinären Planung und produktbezogenen virtuellen Optimierung von automatisierten Produktionssystemen, *Maschinenbau*, vol. 10. Westdeutscher Universitätsverlag GmbH, Bochum (2016)

8. Lee, E.A.: Cyber Physical Systems: Design Challenges. In: 11th IEEE International Symposium on Object and Component-Oriented Real-Time Distributed Computing, pp. 363–369 (2008). DOI 10.1109/ISORC.2008.25
9. Monostori, L.: Cyber-physical Production Systems: Roots, Expectations and R&D Challenges. Procedia CIRP **17**, 9–13 (2014). DOI 10.1016/j.procir.2014.03.115
10. Tolio, T., Ceglarek, D., ElMaraghy, H.A., Fischer, A., Hu, S.J., Laperrière, L., Newman, S.T., Váncza, J.: SPECIES—Co-evolution of products, processes and production systems. CIRP Annals - Manufacturing Technology **59**(2), 672–693 (2010). DOI 10.1016/j.cirp.2010.05.008
11. Ghielmini, G., Pedrazzoli, P., Rovere, D., Terkaj, W., Boër, C.R., Dal Maso, G., Milella, F., Sacco, M.: Virtual Factory Manager for semantic data handling. CIRP Journal of Manufacturing Science and Technology **6**(4), 281–291 (2013). DOI 10.1016/j.cirpj.2013.08.001
12. Devedžić, V., Gaševic, D.: Web 2.0 & Semantic Web, vol. 6. Springer US, Boston, MA (2009). DOI 10.1007/978-1-4419-1219-0
13. Alexander, C., Ishikawa, S., Silverstein, M., Jacobson, M.: A pattern language: Towns, buildings, construction. Oxford University Press, New York, NY (1977)
14. Gamma, E., Helm, R., Johnson, R., Vlissides, J.: Design patterns: Elements of Reusable Object-Oriented Software, 21 edn. Addison-Wesley professional computing series. Addison-Wesley, Boston (2000)
15. International Electrotechnical Commission: Engineering data exchange format for use in industrial automation systems engineering - Automation markup language - Part 1: Architecture and general requirements (2014). IEC 62714-1:2014
16. Weihua, L., Shixian, L.: Improve the semantic interoperability of information. In: 2004 2nd International IEEE Conference on 'Intelligent Systems'. Proceedings (IEEE Cat. No.04EX791), pp. 591–594 (2004). DOI 10.1109/IS.2004.1344818
17. Drath, R., Lüder, A., Peschke, J., Hundt, L.: AutomationML - the glue for seamless automation engineering. In: Factory Automation (ETFA), pp. 616–623 (2008). DOI 10.1109/ETFA.2008.4638461

Behavior Trees for Task-Level Programming of Industrial Robots

Akos Csiszar[1], Martin Hoppe[2], Shahbaz A. Khader[3], and Alexander Verl[4]

[1] GSaME, University of Stuttgart, Stuttgart, Germany
akos.csiszar@gsame.uni-stuttgart.de,
[2] ISW Student, University of Stuttgart, Stuttgart, Germany mjc.hoppe@gmail.com,
[3] ABB Corporate Research, Västerås, Sweeden shahbaz.khader@se.abb.com,
[4] ISW, University of Stuttgart, Stuttgart, Germany verl@isw.uni-stuttgart.de

Abstract. The number of industrial robots used worldwide has been continuously increasing. In almost all cases the application program has the form of a text based source code, which has inherent drawbacks (in terms of complexity and ease of development) when compared to the graphical approaches seen in general purpose software engineering. These graphical programming approaches have not been developed, having industrial robots in mind. In this paper a behavior trees based approach for creating and representing source code for robotic applications is proposed. The software architecture and an experimental implementation of the developed programming method is presented and a validation using a common assembly task is shown.

1 Introduction

The number of industrial robots used worldwide for tasks such as pick and place or welding, has been continuously increasing [3]. Usually, these have an application program in a text based source code to define their behavior. The programmer needs to be aware of the robot's properties and process specifications, thus specialists are needed to commission robotic systems. The (re)commissioning can take several weeks and is costly. Aside from the maintenance costs, this impedes the deployment of industrial robots in small companies and in small lot production [14]. In [12] the example of a robotic welding system is given with eight months of manual programming time and 16 hours process time. Offline programming, aided by CAD models and simulations can reduce the manual programming time drastically. However, it does not change the fact, that the robot's program is representation by textual source code. The growing sector of mass-customized products requires new flexible solutions [13].

The text based source code has major drawbacks (in terms of complexity and ease of development) when compared to visual programming languages (VPL) seen in general purpose software engineering. However, these have not been developed having industrial robots in mind and, therefore, need alterations.

In this paper a novel approach for representing and structuring source code for robotic applications is proposed. The behavior tree modeling, known in software engineering, is altered to cope with the special requirements of industrial robot application software. It is used for task-level programming of robots. A well suited software architecture and a prototypical implementation of the developed programming method is shown and a validation using a common robot task is also presented. The proposed method, enables non-expert users to create and modify programs for industrial robots, without the need for knowing a robot programming language.

2 State of the Art

Task-level is a programming abstraction level, above the skill- and primitive-level [13,15,16,20]. In contrast to the two levels below it, which focus on the robot and where the task becomes nearly a side-effect, task-level programming (TLP) focuses on the impact of the robot on an object [9]. It exclusively uses implicit motion (Insert, Pick, Place), whereas skill- and primitive-level contain explicit motions (MoveL) [9]. Today's robotic systems are nearly all on the primitive level.

Other prosed approaches to robot programming are either text-based systems [9], use different models like UML-derivatives [17], flow-charts [1] as VPL or are gesture and language driven [4, 10]. Those approaches are either not capable of representing a wide range of applications yet or too complex for non-expert users (i.e. not on the task-level).

Behavior trees was originally proposed for behavior design for autonomous systems [6] and were afterwards used in game engines [18] where it is replacing State machines (FSM and HSM) [8]. Behavior trees seem to be expressive enough and to achieve task-level abstractions. It has also already been used for industrial robots in a framework, named CoSTAR [5]. The CoSTAR framework among others has some limitations in readability that prevents its use for large programs. Beyond that, the functionality of those concepts is mostly tailored to meet the requirements of special applications.

By definition [7, 19], behavior trees consist of control-nodes, leaf-nodes and transitions, are executed vertically from top to bottom and back up again. Every node, independent of its type, returns a logic result value to its parent, once its execution is finished. A control-node determines the execution (order) of its child-nodes. Each branch of the behavior tree is closed by a leaf node. Leaf-nodes are the behavior trees actions, changing the behavior of the tree and so the program by their returned result. They can contain any kind of actions to control a robot system. In [2] the use of behavior trees are proposed for tasks carried out by multiple robots (e.g. search and rescue). It focuses on the fault tolerant properties of behavior trees (fallback solution)I. The paper has also an extensive literature review which emphasize the advantages of behavior trees over other methods. Ease of use and the inherent recursive structure are cited as advantages.

In this paper extensions to the definition behavior trees is proposed, covering application scenarios of industrial robots. Especially multitasking and events are the currently missing elements to use behavior trees successfully.

3 Concept and Definition

In order to extend behavior trees for programming industrial robots, new elements (nodes, decorators, etc.) have been developed and implemented. Because this paper is focused on both, concept and implementation, in the following, the implementation of existing behavior tree elements as well as the introduction of novel elements together with their implementation are presented.

3.1 Control Nodes

Control-nodes are non-leaf-nodes, representing the logic that determines the order in which leaf-nodes are executed. The standard definition of behavior tree only contains selector and sequencer as control-nodes.

3.1.1 Selector The selector is part of the standard definition of behavior tree, represented by a question-mark (?). It executes one subbranch after the other from left to right. Once one returns true, it returns true. If all subbranches fail it returns false. The selector can be interpreted as a logical 'or' between its subbranches.

3.1.2 Sequencer The sequencer is also part of the standard definition of behavior trees and is represented by an arrow (→). As the selector it executes all subbranches once after each other. Though it only returns true, if all subbranches succeeded, otherwise false. It logically represents an 'and' operation between all subbranches.

Leaf-nodes contain the actions of a behavior tree. As the name implies, leaf-nodes cannot have any sub-branches or children, but contain the the robots actions. For industrial robots the actions are called instantiated skills. Skills can be all kind of actions a system can perform, from single movements up to complex and smart functions like a vision guided pick. By using skills, instantiated in the environment by the logic that is determined by the control-nodes, tasks are obtained and so task-level programming is achieved.

Multitasking is one of the key elements that is required to successfully deploy robots, especially dual arm robots like ABB's YuMi. Since multitasking is not supported by behavior tree's standard definition, we extended the definition by introducing two special control-nodes.

3.1.3 Parallel Selector The parallel selector, represented by (??), executes all subbranches simultaneously. If at least one subbranch was successful, it returns true, if none was successful it returns false. A use case example is a pick with several arms from one or more boxes. Once one succeeded, the rest of the arms can stop searching and the process can continue.

```
function parallelSelector
    foreach child in children do:
        child.execute in Thread;
```

```
    while true do
        boolean isOneChildstillRunning = false
        foreach child in children do:
            if child has succeeded do:
                children.stop;
                return true;
            if child.status = running do:
                isOneChildstillRunning = true;
        if not isOneChildstillRunning do:
            return false;
```

Algorithm 1.1. Parallel Selector

3.1.4 Parallel Sequencer As the parallel selector, the parallel sequencer, depicted by $\rightrightarrows$, starts the execution of all subbranches simultaneously. However it only returns true, if all subbranches were finished successfully. Otherwise it returns false. The pseudocode of the node is similar to the one above.

Currently existing implementations and publications of behavior trees, lack abstraction and readability. The tree must be entirely displayed to be understandable for the user, since the control nodes do not contain any information of the sub-tree's behavior.

Since behavior trees can become fairly large, navigating inside behavior trees becomes an issue. To reach real task level abstraction in behavior trees, a higher degree of abstraction is indispensable. Therefore labels, collapsing end expanding are introduced. Meaningful labels on control nodes enable the user to understand sub-trees' behavior, without navigating all the way down to their skills (leaf-level). By expanding and collapsing control nodes, the user can concentrate on the region of the tree, he is interested in. The labeling needs to be done manually, during the creation of the tree (i.e during programming). We assume this extra effort to be relatively small, compared to the gain in readability. Especially the modification of an existing behavior tree for other users and novices improves significantly.

3.2 Events

Since behavior trees are not event driven, the proposed extension provides a possibility to respond to events, occurring at any time during the execution. These events are handled by independent behavior trees, one assigned to each event. Furthermore, every event can be en- and disabled. Therefore two special leaf nodes are introduced: En- and Disabler, described in Event-Nodes 3.2.1. Events have priorities which needs to be set. This combines the advantages of state machines (event based transitions) and flowchart (automatic transitions).

3.2.1 Event-Nodes Figure 1a shows a main tree, containing two events. Event 2 has the higher priority than event 1. Event 1 is enabled in region A including region B, while event 2 is only enabled in region B.

Table 1b shows which events can interrupt the execution in which region of the behavior tree. Event 1 can generally interrupt region A. If region B is executed, event 1 can interrupt the execution but can also be interrupted by event 2 in region C, due its higher priority.

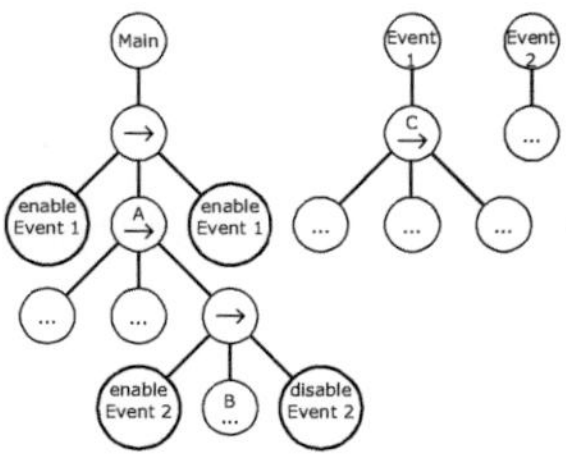

(a) Main and event trees

A	B	C	E1	E2
x			x	x
x	x		x	x
x		x		
x	x	x		x

(b) Interrupt table

Fig. 1. Main tree with two interrupt trees. Event 2 has a higher priority than event 1.

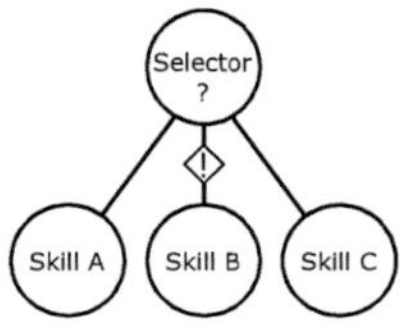

Fig. 2. Not-decorator

```
if skill A succeeds do:
  return true to parent
else if skill B fails do:
  return true to parent
else:
  execute skill c and return
      result
```

Fig. 3. Code representation of figure 2

3.3 Decorators

Decorators are elements, situated on the transition between two nodes of any kind. They always have one parent and one child. Decorators have two capabilities: modification of results and controlling the program flow (resetting sub-trees) These two capabilities are implemented in two different types of decorators. The Invert-Decorator (Inverter) is able to modify results, while the Loop-Decorator is able to re-execute subbranches.

3.3.1 Inverter The inverter changes the result of its child node. This can be useful if a certain behavior is desired when something is not the case. The following small example shows the need:

```
function Inverter
    child.execute;
    if child.result = true do: return false;
    else return true;
```

Algorithm 1.2. Inverter decorator

The description below can be read as $A \vee \neg B \vee C$. This logic term cannot be represented by the logic operations $\wedge$ (and), $\vee$ (or) and the three leaf-nodes without the inverter ($\neg$).

3.3.2 Loop-Decorator The loop decorator is necessary to execute a subbranch more than once. It can either be a for-loop or a while-loop. A for-loop decorator contains four parameters: Start, Count, End, Increment, IgnoreResult. Those parameters can be used globally. Especially the availability of Loop::Count might be interesting for certain applications. The parameters are persistent and retain their values for the entire lifetime of the behavior tree. By that, even notes beyond the decorators' subbranch can always refer to the parameters. IgnoreResult is important to determine if the loop aborts when the subbranch returns returns failure once.

```
function ForLoop
    for Loop::Count from Loop::Start to Loop::End do:
        child.execute
        if not Loop::IgnoreResult && child.result = false do:
  return false;
        Loop::Count increase by Loop::Increment
    return true;
```

Algorithm 1.3. Loop-decorator as for-loop

The while-loop decorator has a different parameter set. As while loops in common programming languages, it requires an expression, that can be validated as a logic value (true / false). A simple example is the consistent execution of a skill until the skill succeeds.

3.4 Data Handling

Like all other representations of industrial robot software, behavior trees have to handle data. We assume all data (variables) to be publicly (globally) available within the application program. To avoid name conflicts, all data has a name, containing the label and the data name, separated by two colons (Pick::PickPosition). If the user does not name nodes properly and a conflict occurs, the link will refer to the first occurrence in the tree. This can only happen, if the user manually labels two nodes the same, since for example two pick skills will automatically be labeled with "Pick1" and "Pick2".

4 Implementation

For the implementation we have chosen an event based approach, combined with the visitor pattern [11]. It consists of two class structures, the model- and the visitor-classes. The model class is the base class for all different behavior tree nodes, the visitor class is obviously for the visitors.

As part of the research an execution visitor was implemented, as modeled in figure 4. It visits the behavior trees' nodes, interprets their content and is able to trigger skills on ABB's YuMi. Those skills are implemented as functions in ABB's robotic language RAPID and triggered via the UDP protocol.

Once the visitor triggers a skill, it waits for the robot to return the result (success, failure) before it returns to the parental control node. The result of each node is displayed in real time on the computer screen by a green (success) and red (failure) dropped shadow. Currently executed nodes are highlighted in blue on the user interface. If the visitor meets a parallel sequencer node, multiple visitor are instantiated and the instantiating visitor waits fro these to return before destroying them. A scenario has been defined to validate the theoretical considerations. Figure 8 in Appendix shows a part of the execution of the validation scenario.

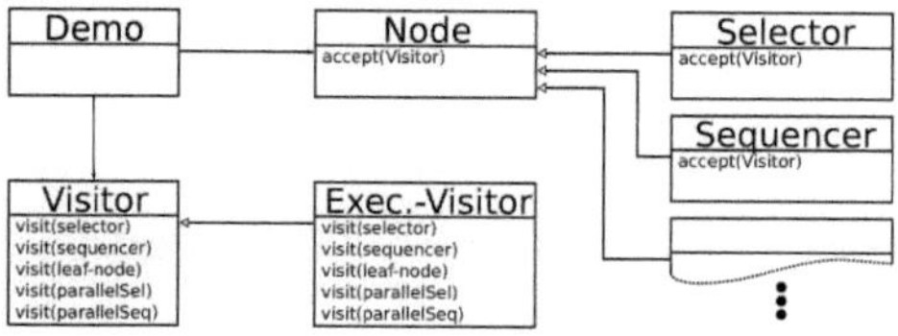

Fig. 4. Visitor pattern scheme

5 Validation

The validation scenario is designed to cover a variety of different use-cases. It is based on Lozano-Pérez' scenario [9]. The task performed is the assembly of a planetary gear carrier (carrier), three planetary gears and three circlips, securing the gears, as shown in figure 5a.

The production line is set-up as shown in figure 5b. The conveyor delivers the planet carriers with an uncertainty. A supply device (FelxFeeder) provids the parts dispersed on a flat surface, provides the belonging planetary gears. A human, standing in front of YuMi robot is responsible for placing the circlips on the shafts to secure the three planet gears.

The robot needs to register the arrival of new planet-carriers by an event from the optical barrier of the conveyor. The robots right hand camera then has to detect the actual position of the arriving part, so that the hand can pick and place it on the buffer. The left hand is supposed to wait for the buffer to contain at least one planet-carrier, to start the assembly. Once a planet-carrier is in the buffer, the left hand picks it subsequently. Simultaneously, the right hand starts to pick the first of the three gears from the FlexFeeder and inserts it on the planet-carriers shaft. The next step includes human interaction. The robot positions its left hand, containing the assembly, towards the human. Once the position is reached, it switches to compliant mode. The human should now be

(a) (b)

Fig. 5. Validation scenario to assemble planet gear carrier (figure 5a) in assembly station (figure 5b) (a) Planet carrier ① with three planet gears ② and circlips ③ (b) Assembly station for planet carrier, containing YuMi Ⓐ, conveyor Ⓑ, ScanFeeder Ⓒ, buffer Ⓓ and tray Ⓔ

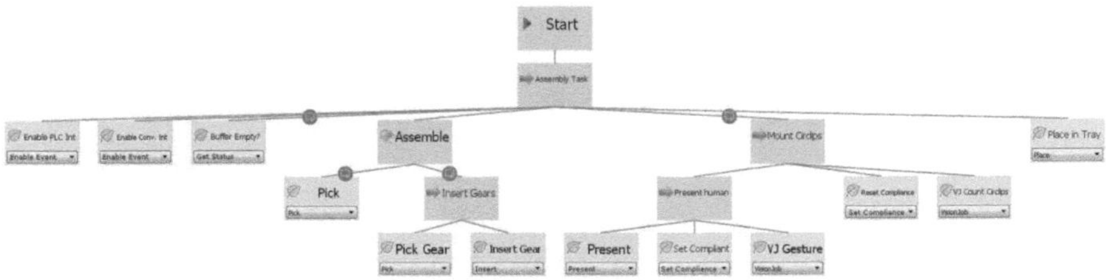

Fig. 6. Main behavior tree representing the evaluation scenario

able to mount the three circlips, while positioning the robots arm as needed. The human indicates, that he has finished by a gesture to the right hands camera. Subsequently the robot must disable its compliance, move to a monitor position and check visually if all circlips were assembled properly with another vision-job. Depending on the vision-jobs result, the robot places the successfully assembled part with the left arm in the tray or otherwise presents it to the human again. If no human is present, the robot drops the current part, into the rejection bin once the input buffer is completely filled and needs to be processed.

The planet-carriers randomly arriving on the conveyor might cause the case, that YuMi is still assembling, while the next planet-carrier arrives. In this case, YuMi has to pause the assembly, put the gear, that might be in its right hand, back onto the FlexFeeder. Subsequently it has to pick the planet-carrier from the conveyor and place it in the buffer, if not full. Once an assembly is finished, YuMi can continue to assemble the accrued planet-carriers from the input buffer.

If the tray is full, it triggers an event (Tray::Full), that needs to pause the entire assembly process and wait for a user input, signaling the replacement of the tray. During this pause, arriving parts should still be placed in the buffer upon their arrival. The periphery is assigned to the behavior tree using a graphical interface and an application modeling approach not in the scope of this paper.

The algorithmic description 7.1 in the Appendix represents the work-flow of the scenario precisely.

Figure 6 shows the main behavior tree of the evaluation scenario. It was designed with our implemented Behavior-Tree-Designer and it is executed using our interpreter based on the visitor pattern, described above. The two event behavior trees below (figure 7) are connected to events raised by the conveyor (7a) and the PLC (7b).

5.1 Results

The behavior tree implemented as proposed in this paper could successfully represent the robots desired behavior. The graphical editor implemented has been used to define the behavior tree and the interpreter, based on the source code fragments from the paper (completed with additional elements like communication to robot and to the user interface) successfully commanded the execution

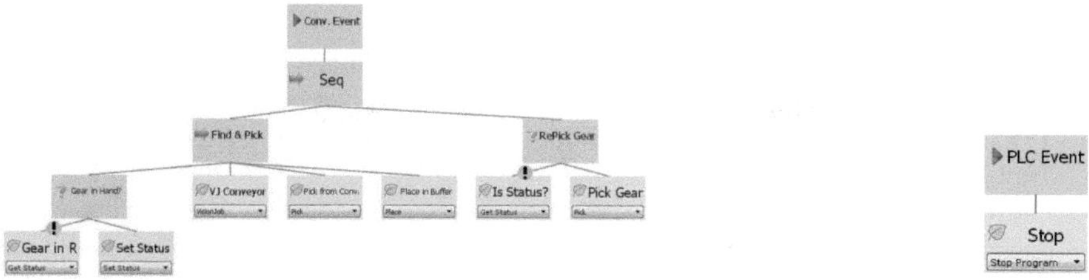

(a) (b)

Fig. 7. Event behavior trees, triggered by the corresponding event from main tree 6

of the task using the robot skills instantiated in the behavior tree. Since this is a use-case for ongoing research, safety aspects have not been considered. Currently available application programming languages do not offer this level abstraction direct comparison to other programming languages is not straightforward.

6 Conclusion and Future Outlook

The behavior tree extension, presented in this paper, based on the validation, is promising. It combines good readability and navigation features by its task-level abstraction and offers a wide functional range to cover many industrial scenarios. At the same time it is easy to learn and understandable almost at first sight.

Nevertheless there are some challenges left to tackle. First of all, currently there is no versatile skill library available [13]. Furthermore, the global data handling should be investigated and improved. Moreover, skills should have an interface definition to make them reusable.

References

1. Sonya Alexandrova, Zachary Tatlock, and Maya Cakmak. Roboflow: A flow-based visual programming language for mobile manipulation tasks. *IEEE International Conference on Robotics and Automation (ICRA)*, pages 5537–5544, May 2015.
2. M. Colledanchise, A. Marzinotto, D. V. Dimarogonas, and P. Oegren. The advantages of using behavior trees in mult-robot systems. In *Proceedings of ISR 2016: 47st International Symposium on Robotics*, pages 1–8, June 2016.
3. IFR Statistical Department and Carsten Heer. World robotics 2016 - industrial robots post a new sales record in 2015 a rise of 8 percent. 2016.
4. Staffan Ekvall and Danica Kragic. Robot learning from demonstration: A task-level planning approach. volume 5, pages 223–234. InTech, 2008.
5. K. R. Guerin, C. Lea, C. Paxton, and G. D. Hager. A framework for end-user instruction of a robot assistant for manufacturing. In *2015 IEEE International Conference on Robotics and Automation (ICRA)*, pages 6167–6174, May 2015.
6. B. Iske and U. Ruckert. A methodology for behaviour design of autonomous systems. In *Intelligent Robots and Systems, 2001. Proceedings. 2001 IEEE/RSJ International Conference on*, volume 1, pages 539–544 vol.1, 2001.

7. Damian Isla. Handling complexity in the halo 2 ai. In *Game developers conference*, volume 12, 2005.
8. Chong-U Lim. An a.i. player for defcon: An evolutionary approach using behavior trees. Master's thesis, Department of Computing Imperial College London, 2009.
9. T. Lozano-Perez. Robot programming. *Proceedings of the IEEE*, 71(7):821–841, July 1983.
10. Pedro Neto, J. Norberto Pires, and António Paulo Moreira. High-level programming and control for industrial robotics: using a hand-held accelerometer-based input device for gesture and posture recognition. *CoRR*, abs/1309.2093, 2013.
11. J. Palsberg and C. B. Jay. The essence of the visitor pattern. In *Computer Software and Applications Conference, 1998. COMPSAC '98. Proceedings. The Twenty-Second Annual International*, pages 9–15, Aug 1998.
12. Zengxi Pan, Joseph Polden, Nathan Larkin, Stephen Van Duin, and John Norrish. Recent progress on programming methods for industrial robots. *Robotics and Computer-Integrated Manufacturing*, 28(2):87 – 94, 2012.
13. M. R. Pedersen, L. Nalpantidis, A. Bobick, and V. Krger. On the integration of hardware-abstracted robot skills for use in industrial scenarios. *Aalborg University Copenhagen*, 2014.
14. Alexander Perzylo, Nikhil Somani, Stefan Profanter, Markus Rickert, and Alois Knoll. Toward efficient robot teach-in and semantic process descriptions for small lot sizes. In *Robotics: Science and Systems (RSS), Workshop on Combining AI Reasoning and Cognitive Science with Robotics, Rome, Italy*, pages 1–7, 2015.
15. F. Rovida and V. Krger. Design and development of a software architecture for autonomous mobile manipulators in industrial environments. In *Industrial Technology (ICIT), 2015 IEEE International Conference on*, pages 3288–3295, March 2015.
16. C. Schou, J.S. Damgaard, S. Bogh, and O. Madsen. Human-robot interface for instructing industrial tasks using kinesthetic teaching. In *Robotics (ISR), 2013 44th International Symposium on*, pages 1–6, Oct 2013.
17. U Thomas, Gerd Hirzinger, Bernhard Rumpe, Christoph Schulze, and Andreas Wortmann. A new skill based robot programming language using uml/p statecharts. *IEEE International Conference on Robotics and Automation (ICRA)*, pages 461–466, May 2013.
18. Jonathan Tremblay. Understanding and evaluating behaviour trees. *McGill University, Modelling, Simulation and Design Lab, Tech. Rep*, 2012.
19. Wikipedia. Behavior tree (artificial intelligence, robotics and control). `https://en.wikipedia.org/wiki/Behavior_tree_(artificial_intelligence,_robotics_and_control)`. [Online; accessed 05-July-2016].
20. Stefan Zei. Manipulation skill for robotic assembly. Master's thesis, Technische Universitt Darmstadt, June 2014.

7 Appendix

7.1 Algorithmic description of validation scenario

```
Main Procedure:
    Enable Event PLC EmergencyStop
    Enable Event Conveyor
    if |Buffer| > 0 do:
        Call Assemble Procedure

Assemble Procedure:
  set gearCount to 0
    Execute threads simultaneosly:
        Thread 1:
            Pick PlanetCarrier with LeftHand from Buffer
            decrement Buffer
        Thread 2:
            while |gearCount| < 3 do
                Pick Gear with RightHand from FlexFeeder
                Insert with RightHand at InsertPosList index
                     gearCount
                increase gearCount
  while VisionJob "CountCirclipJob" with RightHand < 3 do:
    Present with LeftHand at PresentPosition
    Set Compliance LeftHand
    wait for VisionJob "GestureJob" with RightHand to dedect
        doneGesture
    Reset Compliance LeftHand
  Place with LeftHand in Tray

Event by Conveyor:
    if new PlanetCarrier arrives do:
        if Gear in RightHand do: // We d o n t have to care about
             other things than the gear
            Place Gear with RightHand in Flexfeeder
            set rePickGear true
        with VisionJob "ConveyorJob" detect PlanetCarrier on Conveyor
        Pick PlanetCarrier with RightHand from Conveyor
        if rePickGear true do:
            Pick Gear with RightHand from FlexFeeder
        return

Event by PLC:
    Exit programm
```

Algorithm 1.4. Algorithmic description of validation scenario

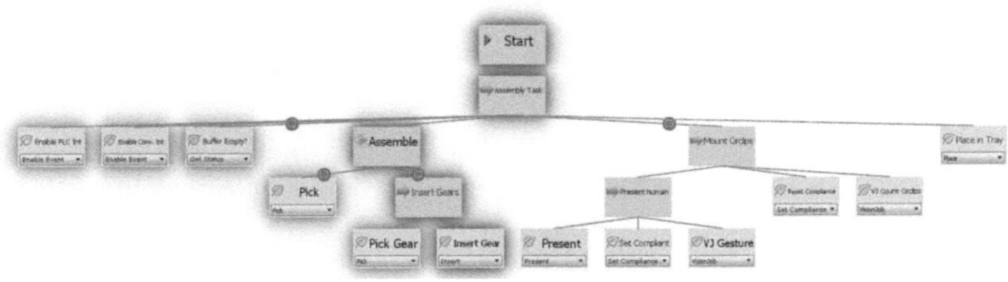

(a) Executing "Insert Gear" for the 3rd time

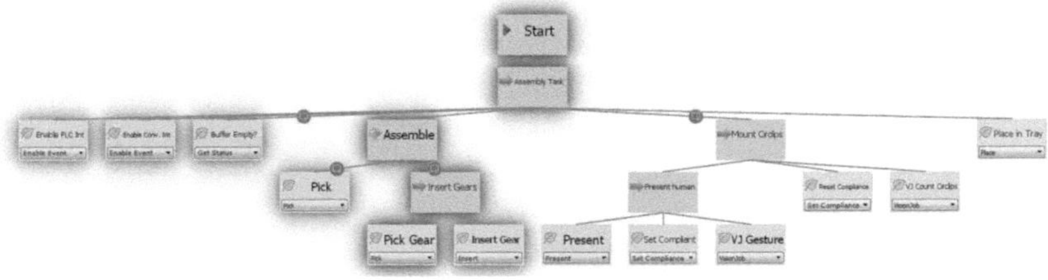

(b) "Insert Gear" failed, so retry until three gears are successfully inserted by picking another gear

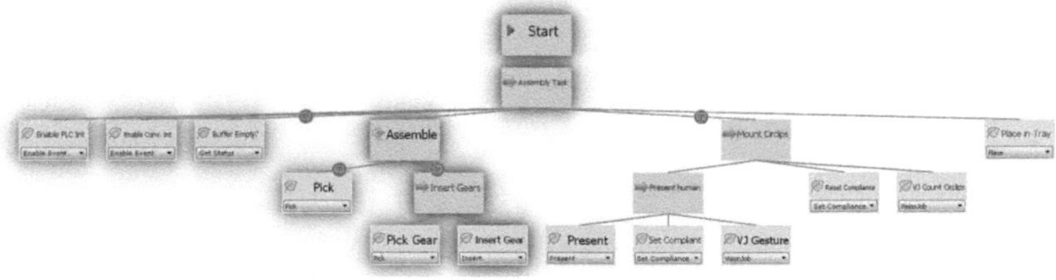

(c) Trying to insert the new gear

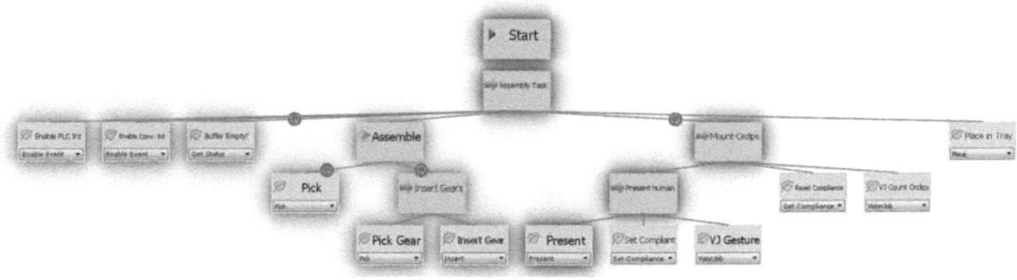

(d) Insertion succeeded (three gears are on the carrier), continue with presenting it to human

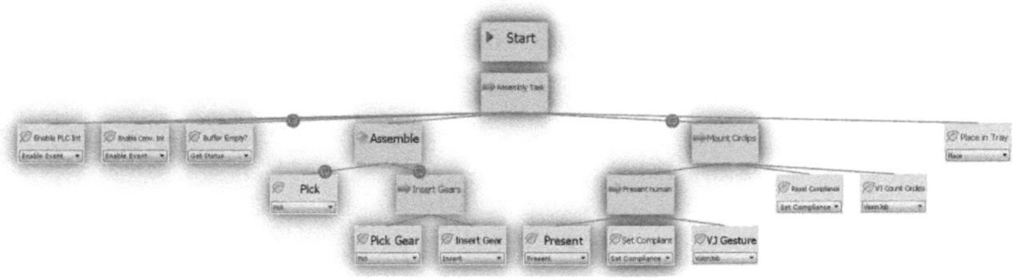

(e) Set hand compliant in "present position", so human can add the three circlips to secure the gears

Fig. 8. Part of the execution of the behavior tree of validation scenario (blue = currently executed, green = succeeded, red = failed)

Work Space Layout and Evaluation Using the Elbedome 2.0 Virtual Environment

Michael Schenk[1*], and Steffen Masik[1]

[1] Fraunhofer Institute for Factory operation and Automation IFF, Magdeburg, Germany
firstname.secondname@iff.fraunhofer.de

Abstract. The paper describes the need for industrial virtual reality (IVR) applications that validate innovative work spaces, which include human-robot collaboration and cognitive assistance systems, factoring in noise, lighting and ergonomics. The capabilities and limitations of two different IVR systems based on head mounted displays (HMD) and mixed reality (MR) are compared. A brief description of the Elbedome 2.0 MR environment is followed by the conclusion and outlook.

Keywords: virtual reality, mixed reality, work space layout, ergonomics

Introduction

In response to manufacturing's growing demands for responsiveness, quality and complexity, Industry 4.0 advocates comprehensive interconnectivity of manufacturing systems, real-time capable digitization of manufacturing processes and performance, and assistance from smart work systems [1, 2]. Designing such future work systems [3] includes such challenges as safe human-robot collaboration and physical and cognitive assistance systems for workers and maintenance technicians. Despite their growing complexity, such work systems nevertheless have to remain responsive to new demands, i.e. be modifiable or upgradable. Such systems, especially in assembly stations, should, however, primarily provide workers effective assistance when they are performing their individual jobs as well as exactly the physical or informational assistance that they need and accept. Not only physical exhaustion but primarily also cognitive and psychological stresses caused by complex, frequently changing and time-sensitive work processes have to be prevented, which can not only lead to drops in performance and quality but also accidents at and absences from work due to stress.

Virtual Reality Work System Design

Shortening planning cycles in Industry 4.0 necessitates evaluating ergonomics, usability and functionality parallel to planning, i.e. even before a work system is really available. Simulation methods, e.g. for ergonomic analysis, can be employed for component parts but involve complex modeling. This is where integral interactive models, so-called industrial virtual reality (IVR) [4], which visually, acoustically and functionally simulate work systems, assembly stations in this case, and enable empirical analyses, can help. Advanced head mounted displays (HMD) and appropriate interactive devices with limited haptic feedback combined with IVR are particularly suited for giving subjects a sense of being present in a work space and being able to perform the assembly process being tested. Movements of the limbs of subjects wearing an HMD have to be scanned by a motion tracking system in real time and assigned to their avatars, i.e. their representation in virtual reality, to enable subjects to sense them (see Fig. 1). Another benefit is that the captured motion data can be used along with standard worksheets (e.g. EAWS) to assess motion sequences.

Fig. 1. Subject with an Oculus Rift DK2 HMD and the target set for IR tracking (left), the subject's avatar in a virtual work space (center) and the subject's view through the HMD (right)

Industrial virtual reality can be used to safely test and evaluate assembly stations with collaborative robot systems (see Fig. 2), maintenance processes in robot cells (see Fig. 3), and other hazardous work processes. Psychological stresses caused by noise, poor lighting or hectic work environments, which are increasingly becoming the focus of attention, can also be simulated and incorporated in an assessment.

Fig. 2. Evaluating a work process at an HRC assembly station with a KUKA LBR iiwa for immobilization

Fig. 3. An interactive HRC work process

Virtual reality based on HMD is nevertheless not free of feedback, especially when stress parameters or process times are being measured and the effect of such interference has to be ascertained by using standardized questionnaires before and after the evaluation. Interviews are usually based on the Igroup Presence Questionnaire (IPQ), a scale for measuring the sense of presence experienced in a virtual environment (VE) [5], and the Simulator Sickness Questionnaire (SSQ), which determines the degree of cybersickness experienced by VE users of [6]. The additional costs and labor for the real-time-capable and sufficiently precise motion tracking systems essential for detailed representations of individuals' avatars are drawbacks. Current HMDs such as HTC's Vive or Facebook's Oculus Rift, on the other hand, only enable capturing transformations of heads and hands in a small range free of visual obstructions. Furthermore, the unspecific haptic and tactile feedback, conveyed by different vibrations through an HMD's hand controller, combined with the absence of weights of the loads is inadequate, especially in assembly processes. Given the more complex technical requirements, latencies of tracking and visualization, and restrictions on natural communication among subjects, work processes performed by several subjects simultaneously can only be reproduced to a limited extent with this method.

Combination with Real Work System Components

At present, standard worksheets as well as empirical-analytical methods based on motion capturing can be used to assess the ergonomics of real work systems. Subjects' ability to receive real haptic feedback and to sense the weights of the loads is an advantage. The late availability of the real work space and environment is a drawback. Simplified prototypes or sets of flexible work space modules can be employed to perform these tests under laboratory conditions parallel to planning (see Fig. 4).

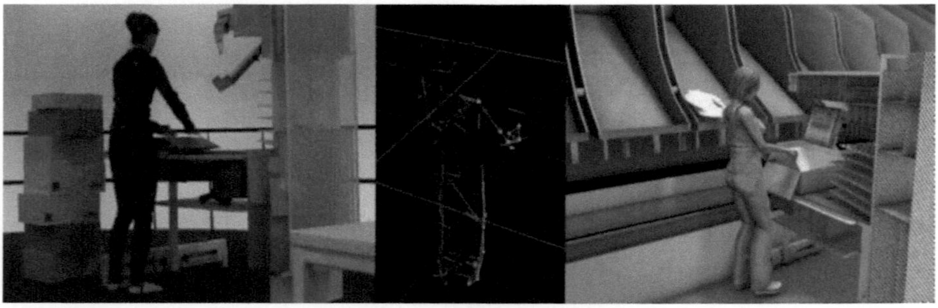

Fig. 4. Work process in a work space prototype (left), biomechanical model (center), and embedding in the planning data (right)

Current research projects are focusing on lessening the drawbacks of methods based on HMDs by linking real work system components with virtual IVR models in one so-called mixed reality. An assembly station is represented by a simplified replica based on a module set or a prototype and ambient conditions such as light, noise and surrounding manufacturing systems and processes simulated by IVR or 360° stereo films. This makes it possible to receive genuine haptic feedback and sense real weights of the loads. Virtual reality can no longer be represented in this case by means of HMD's, though. Instead, this has to be done by stereoscopic projection in the subject's environment. Since subjects are able to see and sense themselves, motion capture can be dispensed with if the kinemetry is analyzed some other way. Although the interference from feedback has to be studied, it is expected to be lower in this setup. A mixed reality system that allows such a setup technically and physically is currently under construction in the Elbedome 2.0 at the Fraunhofer Institute for Factory Operation and Automation IFF in Magdeburg, Germany.

The Elbedome 2.0: An Industry 4.0 Laboratory

The Elbedome 2.0 is the next evolutionary stage and logical continuation of the virtual environment display system [7] opened in 2005 and is focused on the information technology challenges of Industry 4.0. The size and shape of its projection surface immediately set it apart from standard systems used in industry. Its size allows realistic representation of complex and large contents, the presence of many parallel users and the integration of real work systems for mixed applications consisting of real and virtual components. Its cylindrical-hemispherical projection surface measuring 16 meters in diameter allows a continuous omnidirectional stereoscopic representation of virtual contents in a 360° panorama including the floor. Elbedome 2.0 picks up where standard systems reach their limits, thus providing ideal conditions for industrial applications that involve large numbers of people, are highly complex and increasingly require communication.

Elbedome 2.0 was designed to be an Industry 4.0 laboratory and will be available in the future as an innovative tool for research on current issues including not only mixed reality methods for smart work system design and assessment but also monitoring and control of commissioning and operation of Industry 4.0-capable manufacturing systems using cyber-physical systems (CPS), reliable semantic, technical and organizational interoperability (Virtual Fort Knox [8], Industrial Data Space [9]), and sustainable qualification [10] and creativity enhancement in industrial planning.

This will make it possible to us the Elbedome 2.0 in the future as a bidirectional monitoring and control interface to cyber-physical systems created out of von digital models in product and process planning. The heterogeneous models synchronized with the real system have to be converted into a visualizable, interactive and functional industrial virtual reality model and coupled with real-time data from the manufacturing system. This enables representing complex available data manageably and graphically for people to be able to evaluate and verify the complete system's current performance and function. Crucial knowledge is obtained especially whenever current data sets are not only used but also compared with simulated conditions, e.g. operative forecasts or troubleshooting, and, conversely, the real system can be altered. In the use case described here, the interface to the real manufacturing system's CPS makes it possible not only to represent simulated, but also past and present situations in the environment.

The Elbedome 2.0 will provide a space of 50 m² in its center (see Fig. 5), which can be used purely virtually for floor projection as well as for MR scenarios and to design real or prototype work systems. What is more, all important media (power supply, network, compressed air, etc.) are also provided to connect tools, robots and machines inside it. The floor projection system in this area can be used to simulate different lighting situations.

Fig. 5. Elbedome 2.0 with a real work system (sectional view)

Additional elements such as robots' dynamic danger and safety zones, assistance information on individual process steps or interactive menus can also be projected. A work space's environment can be represented omni-directional-stereoscopically on the surrounding projection surfaces, either using a IVR model or 360° films filmed in the real factory. The sound systems delivers realistic sound effects and the integrated infrared tracking system continuously captures subject's movements.

Conclusion and Outlook

Industrial virtual reality makes it possible to experience and evaluate planned work systems at an early stage. Not only can ergonomics be assessed but psychological stresses can also be ascertained by measuring heart rate variability or fatigue by electromyography. Representations in HMDs can entail severe feedback caused by the HMD's weight, cables or laptop backpack, absent natural communication, limited tracking, etc. Moreover, every subject experiences very individual feedback from poor sense of presence and proneness to cybersickness, which, in turn, heavily influences the motion and stress profiles measured. Questionnaires can help classify individual feedback.

A mixed reality approach that lessens the effect of feedback has been proposed. The Elbedome 2.0 is currently under construction and will be open for the first tests in the fourth quarter of 2017. What is more, the findings and results of the evaluation will be published. Future work will involve establishing the feasibility of projections not only on the floor but also on simplified equipment in the center of the Elbedome in order to augment the real components, thus either increasing the level of detail and improving acceptance and usability or enhancing work places with artificial contents to test and evaluate assistance systems.

References

1. Schenk, M. (2014) Industry 4.0-Wege und Lösungsbeispiele. In: Wolfgang Kersten, Koller, H. und Lödding, H. (Ed.): Industry 4.0. Wie intelligente Vernetzung und kognitive Systeme unsere Arbeit verändern. Berlin: Gito (Schriftenreihe der Hochschulgruppe für Arbeits- und Betriebsorganisation e.V. (HAB)).
2. Schenk, M.; Wirth, S.; Müller, E. (2014) Fabrikplanung und Fabrikbetrieb: Methoden für die wandlungsfähige, vernetzte und ressourceneffiziente Fabrik. 2., vollst. überarb. u. erw. Aufl. 2014. Berlin: Springer Berlin.
3. Schenk, M.; Haase, T.; Keller, A.; Berndt, D. (2016) Herausforderungen der Mensch-Technik-Interaktion für die Gestaltung zukünftiger Arbeitssysteme. In: Schlick, C. (Hg.): Megatrend Digitalisierung. Potentiale der Arbeits- und Betriebsorganisation. neue Ausgabe. Berlin: Gito (Schriftenreihe der Wissenschaftliche Gesellschaft für Arbeits- und Betriebsorganisation (WGAB) e.V, 1), S. 131–140.
4. Masik, S.; Schulze, T.; Raab, M.; Lemessi, M. (2016) Comprehensive 3D Visualization of Simulated Processes in Virtual Factories; Proceedings of The 13th International Conference on Modeling, Simulation and Visualization Methods (MSV'16)
5. Schubert, T.W., Friedmann, F. & Regenbrecht, H.T. (1999) Decomposing the sense of presence: Factor analytic insights. Presented at the 2nd International Workshop on Presence, University of Essex, UK, 6-7 April 1999.
6. Kennedy, R.S., Lane, N.E., Berbaum, K.S. & Lilienthal, M.G. (1993) Simulator Sickness Questionnaire: An Enhanced Method for Quantifying Simulator Sickness. The International Journal of Aviation Psychology, 3:3, 203-220, doi: 10.1207/s15327108ijap03033
7. Schoor, W.; Masik, S.; Hofmann, M., Mecke, R., Müller, G. (2007) Elbe Dom: 360 Degree Full Immersive Laser Projection System; Proceedings of Virtual Environments IPT-EGVE 2007. doi: 10.2312/PE/VE2007Short/015-020
8. Landherr, M.; Holtewert, P.; Kuhlmann, T.; Lucke, D.; Bauernhansl, T.(2013) Virtual Fort Knox - Eine flexible und sichere Plattform für Engineering-Anwendungen in Fabrikplanung und –betrieb; wt Werkstattstechnik online 103 (2013), No.2, pp.146-151.
9. Otto, B.; Jürjens, J.; Schon, J.; Auer, S.; Menz, N.; Wenzel, S.; Cirullies, J. (2016) Industrial Data Space – Digitale Souveränität über Daten (White Paper); http://www.industrialdataspace.org/wp-content/uploads/2016/07/Industrial-Data-Space_whitepaper.pdf.
10. Haase, T.; Weisenburger, N.; Termath, W.; Frosch, U.; Bergmann, D.; Dick, M. (2014) The Didactical Design of Virtual Reality Based Learning Environments for Maintenance Technicians; Virtual, Augmented and Mixed Reality. Applications of Virtual and Augmented Reality, 27-38, Springer.

Automated handling of high-temperature thermoplastic Carbon Fiber materials for a large-scale production

Johannes Graf[1*], Christoph Richter[1] and Gunther Reinhart[1]

Fraunhofer Research Institution for Casting, Composite and Processing Technology IGCV, Augsburg, Germany

Johannes.Graf@igcv.fraunhofer.de

Abstract. The structural and mechanical advantages of Carbon fiber reinforced composites (CFRC) lead to an increasing demand of Carbon fiber products. This class of materials is gaining widespread acceptance in various fields like aviation, wind energy or especially automotive and is gradually replacing traditional lightweight construction materials such as high-strength steel or aluminum. Currently, particular process steps of the production of fiber composite structures are performed manually or semi-automatically. Especially the automated handling of semi-finished products consisting of unstable textile poses a challenge for an economical manufacturing. Due to the manifold advantages of thermoplastic composites and the challenges concerning the processing, the paper presents the conception for the automated handling of high-temperature thermoplastic semi-finished products.

Keywords: Thermoplastic Carbon Fiber Composites; Automation; Handling;

1 Introduction

Current discussions on energy and resource efficiency have significantly influenced the importance of Carbon fiber reinforced composites (CFRC). Additionally, present social trends such as sustainability highlight FRP in the current research landscape. The rising desire for low-emission and cost efficient mobility in the future emerges in novel challenges in the areas of electrification and lightweight. Thereby, FRP components illustrate a high potential for the implementation of lightweight structures [1, 2]. Especially its construction and the anisotropic structure provide technical advantages, such as lightweight construction, over existing materials in production technology. Further, the economical use of FRP depends on this construction and optimal arrangement of the load-transmitting fibers in the component. Specifically, the use of a thermoplastic matrix initializes new approaches and solutions for a large-scale production. Using an unidirectional fiber structure (e.g. unidirectional composite plates) as a starting material constitutes a key lever, as the fibers can be placed precisely in the fiber orientation to contribute an optimum load [3]. Fig. 1 illustrates a typical industrial process using a unidirectional composite structure.

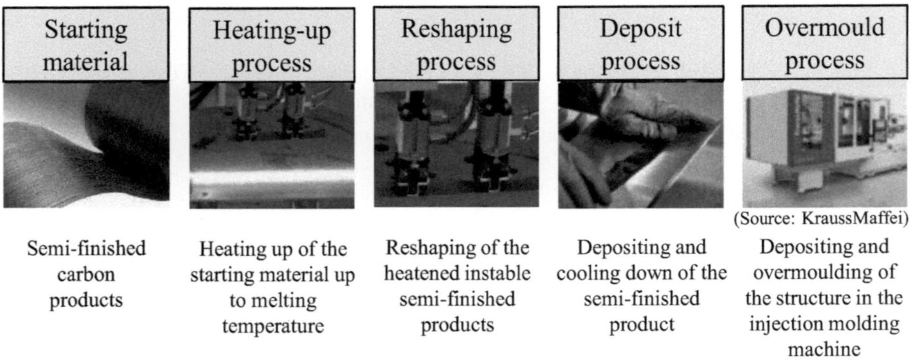

Fig. 1. Typical industrial process chain for the fabrication of thermoplastic FRP [3]

The thermoplastic semi-finished carbon structure is generally compressed from single unidirectional tapes. In the first process step the planar sheet is cut into the desired contour, then transferred to a heating station and heated up to the desired forming temperature. In the next step, the cut material is formed accordingly to the required load path, and finally placed in a laying mold [4]. Furthermore, the individual formed composite structures are joined to a basic structure by a welding process. In addition, metallic inserts can be used for introducing the load. In the final step, the basic structure is placed in an injection molding machine and sprayed with a short-fiber material to the final structure. Especially the reshaping and depositing process steps require high demands on the technical feasibility for an economical automated solution. The paper presents and focuses on a technical solution for the handling and the forming of high-temperature thermoplastic Carbon Fiber materials.

2 State of the Art

In literature numerous approaches for the automated handling of semi-finished components can be found. Thereby, the processing of form-instable dry fiber textiles is focused.

As an example, the handling and stacking of flat, limb materials are focused in [5–8].

Solutions for the automated preforming and draping are brought into focus in [9–13]. Transferred to the fabrication of thermoplastic products, most of the available systems are dedicated to a special application.

[14, 15] describe the conception, selection and detailing of thermoplastic process chains. [16, 17] represent automated fiber placement technologies with a unidirectional thermoplastic fiber material. This technology is based on the repetitive reshaping and depositing of individual tapes for the construction of semi-finished products.

An alternative approach pursue the cost-effective fabrication of unidirectional sheets with the final thickness, based on thermoplastic tapes (e.g. on a double belt press). Concerning the reshaping and draping of these elongated unidirectional plates, there does not exist satisfying automated handling solutions.

3 Conception of an automated handling system

Based on the illustrated process of the handling of heated thermoplastic semi-finished materials, a systematic deductive approach is presented with the aim of developing and constructing a robot based handling system. Due to the different effects of the material for the processing (different deformability and deformation forces), the specification plays a central role for the handling system. Starting with the definition of the used material, Table 1 describes the characteristics of the thermoplastic material.

Table 1. Characterization of the thermoplastic Carbon Fiber Composites material

Parameter	Wert
Thickness	0,7 mm – 3,0 mm
Length	100 mm – 600 mm
Width (b)	30 mm
Heating mechanism	Infrared oven
Thermoplastic material	PA (Polyamid) 6
Fiber type	Sigrafil C
Maximum heating temperature	240° C

First of all, for the definition of the system boundary, the prevailing framework conditions of the experimental environment and the investigations are determined. Therefore, the dimensions of the handling system are predefined due to the measurement of the plates. Further attributes, as the precise material specification (the type of thermoplastic material, the material thickness and the fiber type) and the process parameters (the heating temperature) have to be defined as well.

The exact positioning of the unidirectional sheets has a great influence on the strength of the subsequent component. For example, a fiber deflection of five degrees already results in a reduced loadability of approximately 30% in various applications. For this reason, adherence to the quality requirements frames all levels and is taken into account in the selection process.

Thermoplastic composite structures with a unidirectional fiber structure (width 30,0 mm, thickness 1,30 mm) should be absorbed in the original cooled state and heated above an infrared field. A one-sided heating is simplified. For thicker structures, double-sided heating is used as standard. In the deposited condition, the material is supposed to be deformed in the required position, dimensionally stable in the mold.
In addition, the material characteristics have a significant influence on the design of the handling process. The approach is divided into three levels, which has to be traversed to solve the problem successively. These are the following steps:

- Evaluation of different draping strategies
- Selection of a suitable gripping principle
- Kinematics for the realization of the forming movement

Based on the formulated requirements, these three steps are proposed in the following.

3.1 Evaluation of different draping strategies

Initially, it is essential to analyze the potential types of deformation of the handled material. For a load path orientated construction, the fiber structure has to be formed exactly according to the flow of force. Figure 2 demonstrates the potential forming approaches.

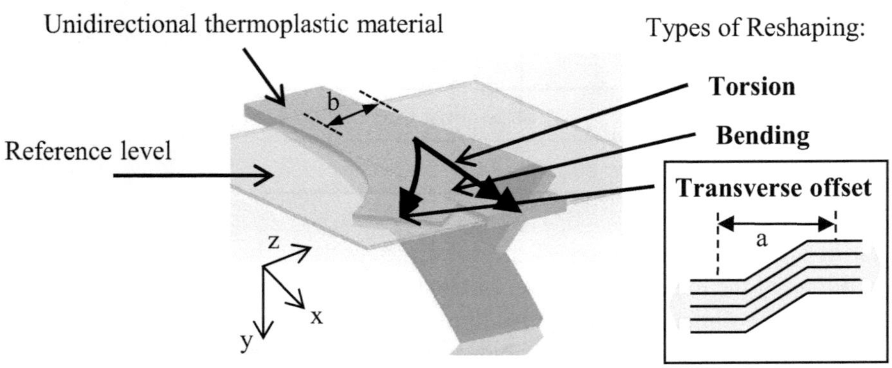

Fig. 2. Types of reshaping for unidirectional thermoplastic carbon materials.

In addition to a curvature (or bending) and to torsion, above all a transverse movement of the material in the feasibility level imposes high requirements on compliance with the position tolerances.
Due to missing flexibility of the fibers, but also a simultaneous fiber tensioning over the entire width of the material, a controlled reshaping has to be ensured.
The transverse offset in z-direction has to be considered separately, since the individual fibers can neither be compressed nor stretched. The concept in the right part of figure 2 illustrates an approach of carrying out a discrete sectional displacement, whereby the fibers remain in tension, but curvatures are possible over a longer section. For this purpose, the material is separately segmented at a distance of several gripping points. The further transformations and draping modes can be depicted in the y-direction and z-direction by linear displacement of the individual gripping points.

3.2 Selection of a suitable gripping principle

Based on the results of the draping strategy and the required types of reshaping of the semi-finished material, a suitable gripping principle for segmented handling has to be selected. Figure 3 shows potential principles, separated into the types of contacting, force fit, tight fit and adhesive joints.

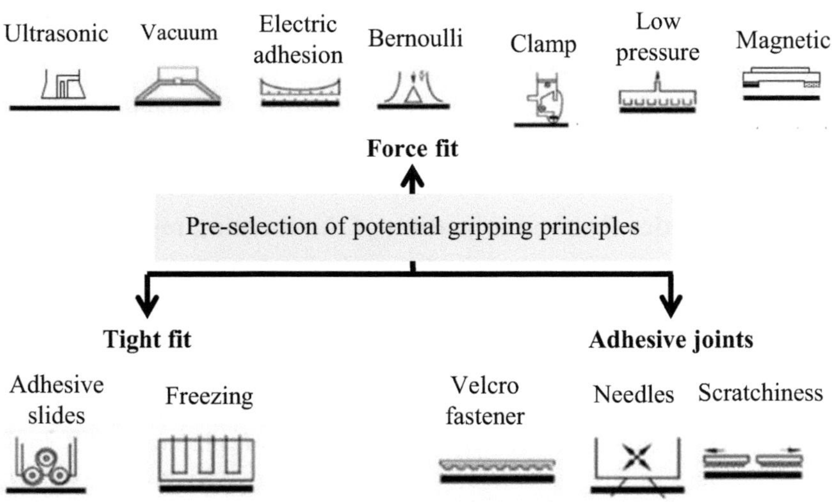

Fig. 3. Overview of different types of contacting with potential gripping principles [18]

Due to the form-stable structure in the initial state and the quality requirements for the further processing, the connection is limited to the force fit. Considering the material characteristics and surface of the thermoplastic semi-finished material (exclusion of ultrasound, electrical adhesion and magnet), the process control (reject of clamp), using a vacuum is the most encouraging principle. As a result of the air flow, it is necessary to check the temperature on the surface of the material. During the handling process the semi-finished material has to keep a sufficient heat capacity, as well as the cooling cannot increase dimensionally.

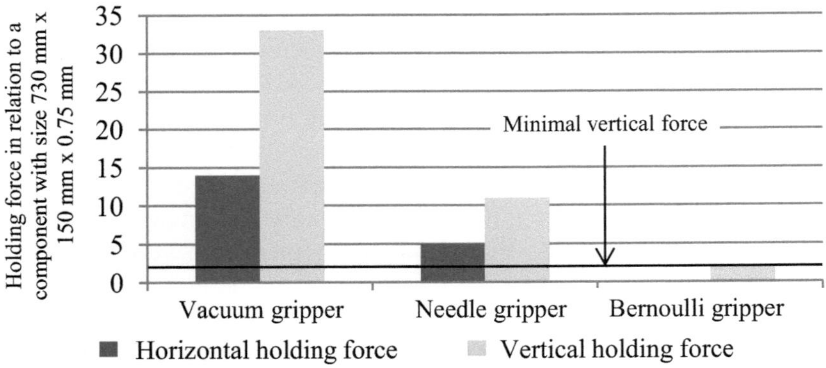

Fig. 4. Different holding forces depending on the potential gripping principles

Figure 4 illustrates the holding forces for the basically suitable principles. With the Bernoulli gripper, the force for a transfer offset cannot be transmitted due to the initial mechanism. However, these forces are necessary for a defined shaping. In addition, the two-dimensional vacuum suction is excluded due to the partial sectional handling.

Adapted to the semi-finished material, temperature resistant vacuum grippers with a diameter $d < b$ are selected for the application, as it covers the complete gripper surface. The theoretical selection is validated on the experimental test series.

3.3 Kinematics for the realization of the forming movement

In the last step, the partial results are summarized in the conception and in the design of a handling system. Therefore, the kinematic design of the gripping points illustrates the basis for the gripper. Figure 5 shows the prototypical concept in the left part. Additionally to the linear displacement in the z-direction for the transverse offset, the suction elements can also be actively displaced linearly in the y-direction for the bending. If torsion is required, a further rotational axis can optionally be implemented. Figure 5 illustrates the concept, exemplified for three elements.

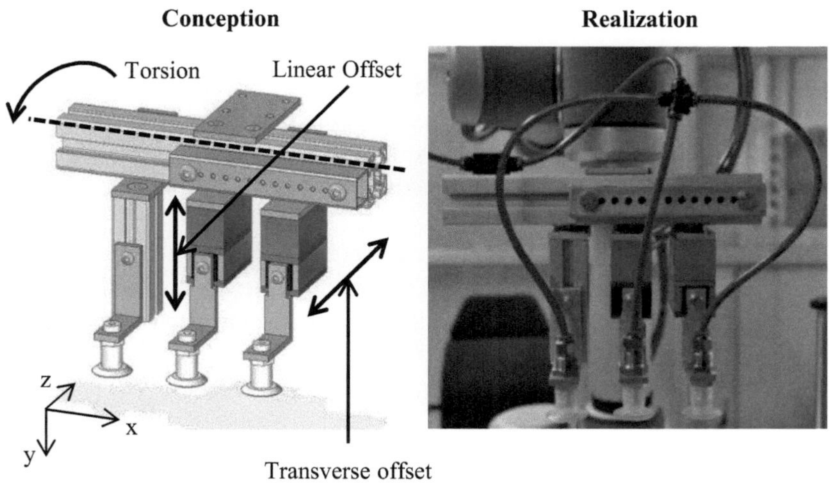

Fig. 5. Conception and realization of the prototypical handling system

4 Use Case and implementation

For a subsequent validation and technical testing of the handling system, the gripper is prototypically constructed and assembled. Figure 5 describes in the right section the finalized gripper. Due to the small weight of the semi-finished material, but also due to the required flexibility for the handling and forming operations, a 6-axis articulated robot with a maximum load of 10 kg is selected. The linear displacements are accomplished with pneumatic units. In order to prove the concept, experimental studies concerning the technical feasibility of the handling system are carried out to investigate the following different criteria. Exemplary results are summarized in figure 6.

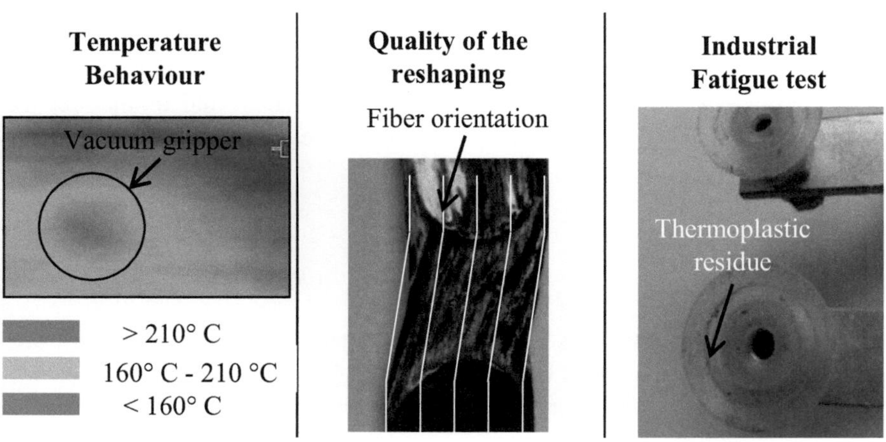

Fig. 6. Experimental studies and results based on different criteria

1. **Temperature behavior**
   Due to the vacuum with a resulting air stream during the suction of the semi-finished material, a temperature change occurs on the upper side. Based on test series with infrared measurements, a sufficient heat capacity can be derived at a thickness of the material with more than 0,7 mm. With this setting, a permanent active shaping is possible.
2. **Quality of the reshaping**
   In order to realize the transverse offset, it is necessary to guarantee all fibers stretched along the flow of the force. Experimental and by means of a micrograph, it can be shown that the fiber orientation takes the desired geometry for this case.
3. **Fatigue test**
   Accordingly to a large-scale production, different sub processes are analyzed for the industrial availability. With regard to this industrial utilization and capacity, the material fatigue strength is investigated. Especially the contact between the vacuum gripper and the thermoplastic material as well as the pneumatics is checked regarding the long-term serviceability. On account of the occurrence of thermoplastic material residues on the surface, a stripping unit is to be considered for this purpose.

5 Summary and outlook

The high costs in the production of composite structures arise largely due to the high share of manual handling operations. In particular, there exist only insufficient automation solutions for the reshaping of unidirectional thermoplastic plates. In the present paper a handling concept was developed and implemented prototypically based on a material specification.

In the next step, the handling system is validated in further test series due to its requirements of the achievable availability as well as due to an economical evaluation. Thereby, the achievable cycle times and the final deposit quality is focused in order to obtain a classification into the entire process chain and to subtract the handling solution from an economic and technological evaluation.

References

1. M. Neitzel, P. Mitschang, and U. Breuer, "Manual composites, materials, processing, application", Carl Hanser, Munich, 2014.
2. Acmite Market Intelligence; "Global Carbon Fiber Composite Market", Watwill, Switzerland, 2014.

3. R. Heuss, "Lightweight, heavy impact; How carbon fiber and other lightweight materials will develop across industries and specifically in automotive"; Advanced Industries; McKinsey & Company, 2012.
4. H. Schuermann, "Designing with fiber-plastic composites", Springer, Berlin, 2005.
5. Lien TK, Davis PG (2008) "A novel gripper for limb materials based on lateral Coanda ejectors". Ann CIRP 57:33–36
6. Saadat M, Nan P (2002) "Industrial applications of automatic manipulation of flexible materials". Ind Robot Int J 29(5):434–443
7. Seliger G, Szimmat F, Niemeier J, Stephan J (2003) "Automated handling of non-rigid parts". Ann CIRP 52(1):21–24
8. G. Reinhart, G. Strasser, "Flexible gripping technology for the automated handling of limp technical textiles in composites industry". Production Engineering 5, pp. 301-306, 2011.
9. G. Reinhart, C Ehinger, "Robot-based automation system for the flexible preforming of single-layer cut-outs in composite industry", German Academic Society for Production Engineering WGP, Heidelberg, 2014.
10. Großmann K, Muehl A, Loeser M, Cherif C, Hoffmann G, Torun A (2010) "New solutions for the manufacturing of spacer preforms for thermoplastic textile-reinforced lightweight structures". Prod Eng 4:589–597
11. Kordi MT, Huesing M, Corves B (2007) "Development of a multifunctional robot end-effector system for automated manufacture of textile preforms". (Hrsg.): IEEE/Asme international conference on advanced intelligent mechatronics: IEEE 2008. ISBN: 978-1-4244-1263-1
12. Greb C, Schnabel A, Gries T, Kruse F (2010) "Development of new preforming processes for high performance fibre-reinforced plastic (FRP) components". Sampe J 5:42–51
13. Brecher C, Emonts M (2011) „Automatisierte Handhabung für die FVK-Großserienproduktion". In: 17. Nationales Symposium SAMPE Deutschland e. V. (Hrsg.): Faserverbundwerkstoffe 2011
14. P. Giliard, J. Graf, M. Jelinek, „Efficient production of fiber composite components by Automation", Internationales Forum Mechatronik, IMS Institut für Mechatronische Systeme, Winterthur, 2013.
15. J. Naegele, J. Graf, G. Reinhart, „Bewertung von Prozessketten der thermoplastischen CFK", Leightweight Design, Volume 8, Issue 6, pp 50–55, Konstruktion und Entwicklung, 2015
16. Dieffenbacher GmbH, „Maschinen- und Anlagenbau, Fiberforge RELAY Station", http://www.dieffenbacher.de/upload/teaser/Flyer/, Eppingen, 2014, Lightweighting your world
17. Coriolis Composites, "Robotized fiber placement technology", http://www.coriolis-composites.com/products.html, Quéven, 2017
18. Hesse, S.: Greifertechnik; „Effektoren für Roboter und Automaten". München: Hanser, Carl. 2011. 300 S. ISBN: 978-3-446-42422-7.

MIX
Papier aus verantwortungsvollen Quellen
Paper from responsible sources
FSC® C105338

If you have any concerns about our products,
you can contact us on
ProductSafety@springernature.com

In case Publisher is established outside the EU,
the EU authorized representative is:
Springer Nature Customer Service Center GmbH
Europaplatz 3, 69115 Heidelberg, Germany

Printed by Libri Plureos GmbH
in Hamburg, Germany